MW00666908

the bird book

DK

Smithsonian

the bird book

THE STORIES, SCIENCE, AND HISTORY OF BIRDS

Senior Editor Gill Pitts
Project Editors Thomas Booth, Becky Gee, Miezan van Zyl
Editors Annie Moss, Hannah Westlake, Michael Clark
US Editor Jill Hamilton
Senior DTP Designer Harish Aggarwal
Production Editor Andy Hilliard
Senior Production Controller Meskerem Berhane
Managing Editor Angeles Gavira Guerrero
Publishing Directors Jonathan Metcalf, Liz Wheeler
Managing Director Liz Gough

Senior Art Editor Sharon Spencer
Project Art Editors Amy Child, Phil Gamble
Design Assistant Noor Ali
Picture Researcher Deepak Negi
Senior Jackets Editorial Coordinator Priyanka Sharma Saddi
Senior Jackets Designer Surabhi Wadhwa-Gandhi
Jackets Design Development Manager Sophia MTT
Managing Art Editor Michael Duffy
Art Directors Maxine Pedliham, Karen Self
Design Director Phil Ormerod

This book was made with Forest Stewardship Council™ certified paper – one small step in DK's commitment to a sustainable future.
Learn more at **www.dk.com/uk/information/sustainability**

First American Edition, 2024
Published in the United States by DK Publishing,
a division of Penguin Random House LLC
1745 Broadway, 20th Floor, New York, NY 10019

24 25 26 27 28 10 9 8 7 6 5 4 3 2 1

001–336922–Sept/2024

Published in Great Britain by Dorling Kindersley Limited

A catalog record for this book is available from the Library of Congress.

ISBN 978-0-5938-4406-9

DK books are available at special discounts when purchased in bulk for sales promotions, premiums, fund-raising, or educational use. For details, contact: DK Publishing Special Markets, 1745 Broadway, 20th Floor, New York, NY 10019
SpecialSales@dk.com

Printed and bound in China

www.dk.com

Contributors

Rob Hume (lead author) is a writer, editor, artist, and identification expert, highly adept at penning detailed identification notes for rarity watchers yet also skilled at providing interesting descriptions for armchair birdwatchers. He is a lifelong birdwatcher who worked for the RSPB for more than 30 years and edited the RSPB's award-winning *Birds* magazine.

Dominic Couzens is one of Britain's most prolific writers on birds and wildlife, having published almost 50 books and almost 1,000 articles in multiple periodicals. He is especially passionate about relaying fascinating and unusual facts about familiar wildlife. He received the British Trust for Ornithology's Dilys Breese Medal in 2021 for science communication.

Chris Harbard is a lifelong birder who has traveled the seven continents looking for birds. Working for the RSPB for many years, he now writes for *Birdwatch* magazine, lectures on expedition cruise ships, and leads birding tours. He lives in Arizona, US, and together with his wife helps run the Southwest Wings Birding and Nature Festival.

James Lowen has been immersed in all aspects of natural history since a common buzzard caught his eye as a toddler. He has fledged into an author specializing in wildlife and travel, with two of his 15 books winning the accolade of Travel Guidebook of the Year.

Josh Jones is an avid birdwatcher and writer based in Lincolnshire, UK. He is editor of *Birdwatch* magazine and the website *BirdGuides.com*, and has a keen interest in bird migration and vagrancy.

Chris Sharpe has worked on Neotropical bird conservation since 1988, based mainly in Venezuela, where he is a Research Associate of the Phelps Ornithological Collection. He has contributed to various journals and to *BirdsoftheWorld.org*. As an ornithologist he runs threatened species (Red Data) assessments, bird surveys, rapid biodiversity inventories, and training workshops—especially on shorebirds—from Alaska to Antarctica.

Marianne Taylor is a lifelong birder and wildlife enthusiast who has worked since 2007 as a freelance writer, editor, illustrator, and photographer. A former editor at the publisher Bloomsbury, she has written more than 30 books on natural history, in particular birds and insects, and regularly works in collaboration with the RSPB.

contents

CHAPTER 3

Passerines

Net gain
This detail of an ancient Egyptian wildfowling scene on the c.3,900-year-old tomb of Khnumhotep II at Beni Hassan depicts various ducks and geese caught in a net. Other birds, including a hoopoe and some shrikes, are shown perched in two acacia trees flowering on a bank along the Nile.

CHAPTER 1

Introducing Birds

Birds are warm-blooded, feathered animals that reproduce by laying eggs. This chapter explains how birds evolved, how they fly, and how they sense and use the world around them. And it highlights their long cultural history with humans, in science, religion, mythology, and art.

What is a Bird?

Like reptiles, birds lay eggs that have a protective shell, and like mammals, they are warm blooded. Uniquely, they are the only animals alive today that have feathers.

Evolution of birds

Some carnivorous dinosaurs became small, walked on two legs, and developed feathers. A few could fly weakly, perhaps feeding in trees. Their fossils are difficult to separate from early birds, such as *Archaeopteryx*, which appeared around 150 million years ago (MYA). Modern birds appeared around 67 MYA.

▲ Birdlike dinosaur
Anchiornis huxleyi predates *Archaeopteryx* by about 10 million years. Its feathers are birdlike, but features such as the bony tail and toothed jaws remain reptilian. Large wings and sophisticated feather shapes suggest that it could glide for short distances rather than simply fluttering above ground as it ran.

EVOLUTION OF FEATHERS

Feathers began as filaments or tufted structures that may have kept dinosaurs warm. Later, some feathers developed a stiff shaft, or rachis. Some of these feathers had narrow barbs with numerous tiny, interlocking barbules knitted into a smooth, flat surface, known as a vane. Only those feathers with a narrow outer vane had the right aerodynamics for flight.

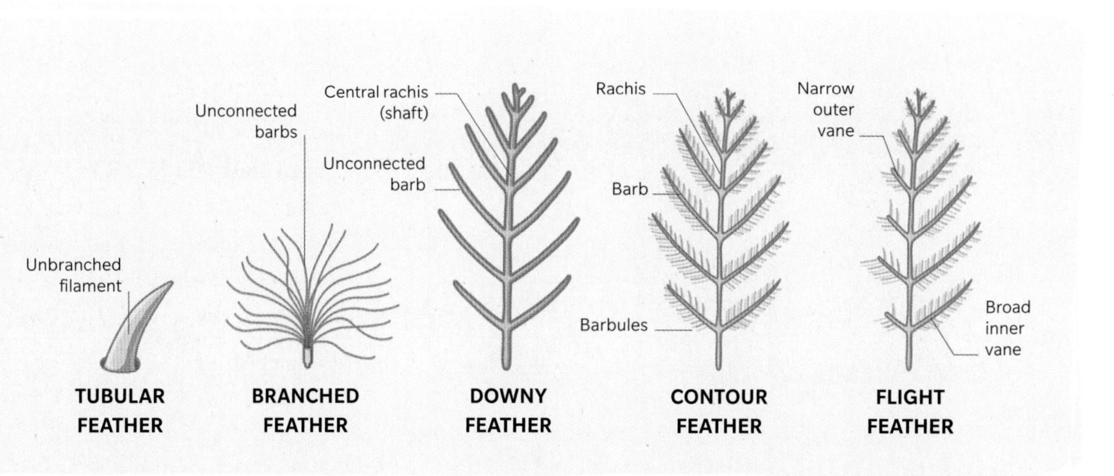

The skeleton of most flying birds weighs less than their feathers

Skeleton

A bird's skeleton is made up of thin-walled but strong bones filled with air cavities. Birds have fewer bones than reptiles or mammals, and many of them are reduced in size and even fused together, such as in the wrist and digital bones of the wing. A deep sternum allows purchase for powerful muscles to beat the wings. Unlike other vertebrates, many bird species can move both their upper and lower mandibles (jawbones).

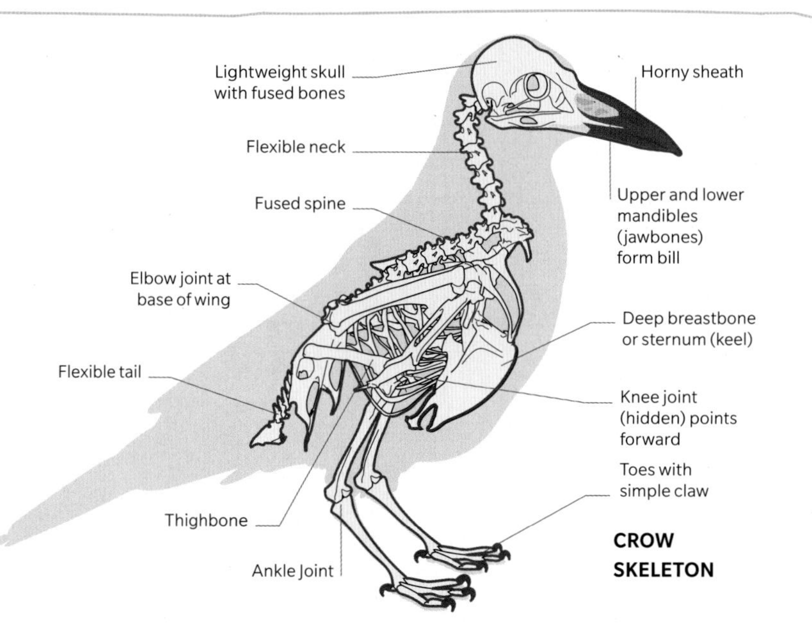

➤ Power and grace
The Bald Eagle (*Haliaeetus leucocephalus*) has a 7.5 ft (2.3 m) wingspan, weighs a muscular 7–13 lb (3–6 kg), yet can soar, glide, and pick up food with its feet from the surface of a lake with great precision.

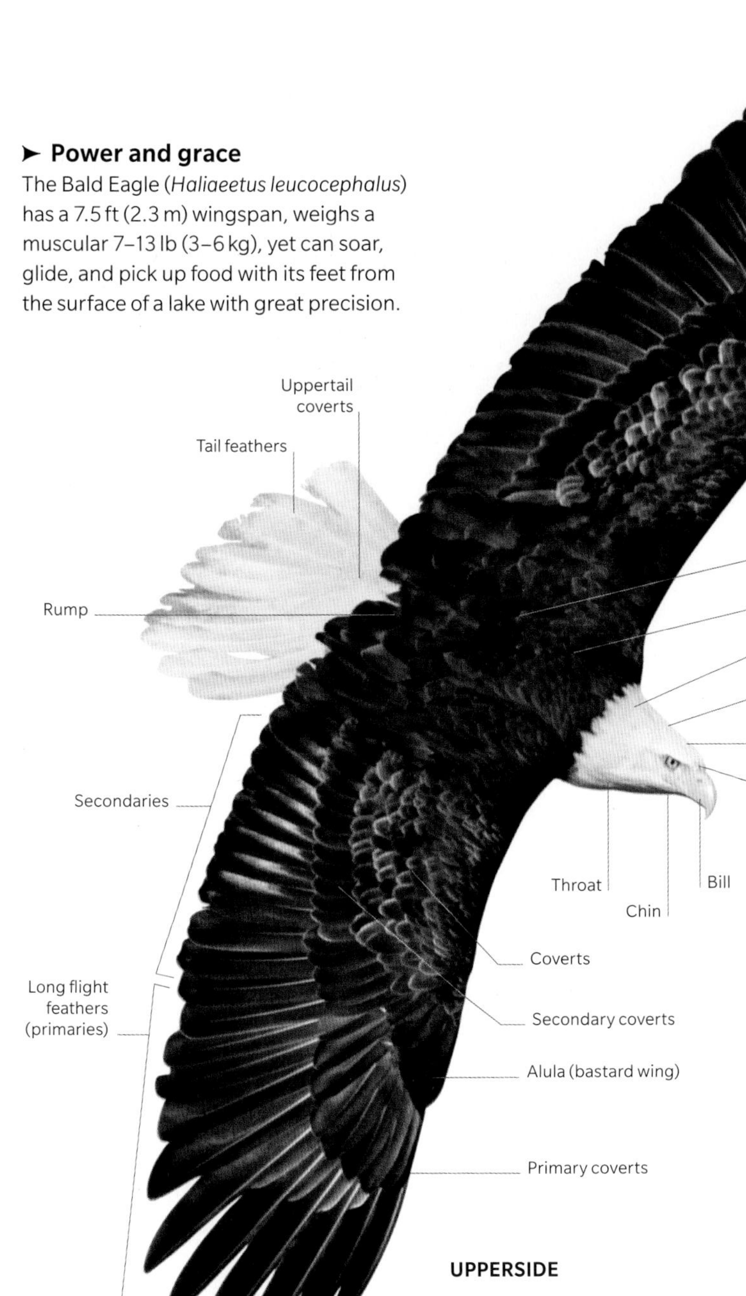

Undertail coverts
Axillaries
Breast
Belly
Vent
Foot
UNDERSIDE

External anatomy

Feathers grow in distinct tracts, which are especially obvious on the wings and back. The head and body are smoothly shaped by contour feathers. Rows of small coverts, each row overlying the base of the next, give a smooth, aerodynamic surface to the wing. The inner part of the wing has large secondary feathers; the wingtip has long primary feathers. On this eagle, the outermost primaries have stepped vanes, creating a series of fingers when spread, reducing turbulence at the wingtip.

Respiration

Birds have relatively small lungs for their body size, but this is compensated by nine additional air sacs positioned at different points throughout their body, so the volume of their respiratory system is 3–5 times that of a mammal. The air sacs continuously feed fresh air directly to a bird's lungs, which means birds can take in oxygen while both inhaling and exhaling.

➤ Air sacs
Airflow from the air sacs is unidirectional, so only fresh air enters the lungs, creating a higher amount of oxygen to diffuse into the bloodstream.

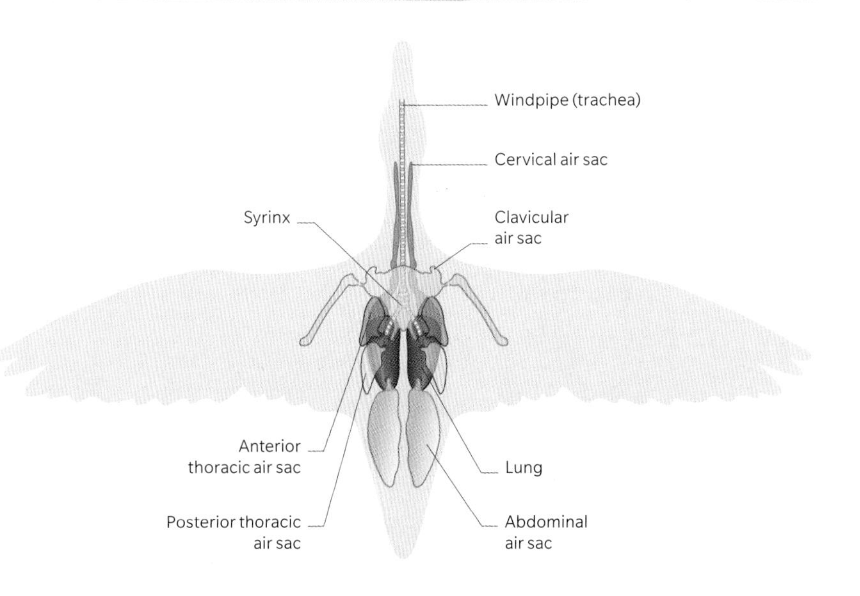

A Brief History of Birds

Early humans hunted wild birds for their meat, eggs, and feathers. The scientific study of birds began in ancient Greece 2,300 years ago, although domestication had begun around 6,000 years previously. Over the millennia, our understanding of their evolution, anatomy, and behavior has changed with every new discovery.

RED JUNGLEFOWL

c.6000 BCE
Chickens were first domesticated from the red junglefowl in Southeast Asia by 6000 BCE, then separately in India by 2000 BCE. Geese are thought to have been domesticated in China by 5000 BCE

2nd millennium BCE
The Vedas, a collection of Hindu religious texts, describes the Asian koel (a member of the cuckoo family) as *Anya-Vapa*, meaning "raised by others." This is the first reference to brood parasitism, where females lay their eggs in the nests of other species.

ATHENIAN COIN

c.350 BCE
In his *History of Animals*, which he wrote in the city-state of Athens, the Greek philosopher Aristotle established that all members of the group "birds" have two bony legs, feathers, wings, and a beak.

SHELDUCK MOSAIC, POMPEI

c.120 BCE
The mallard was domesticated for its meat, eggs, and feathers in China around 500 BCE, whereas the ancient Romans still hunted wild ducks and took their eggs to raise tame ones.

c.137 CE
Cygnus, or The Swan, is one of 48 constellations described by Alexandrian astronomer and astrologer Ptolemy in his *Almagest*. The nine linked stars depict a swan in flight across the Milky Way.

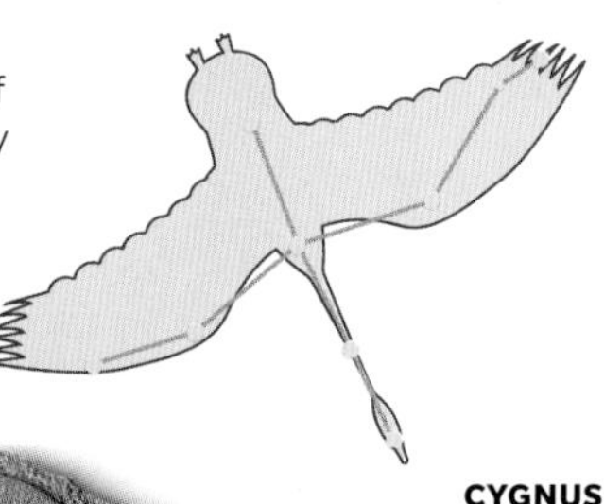

CYGNUS

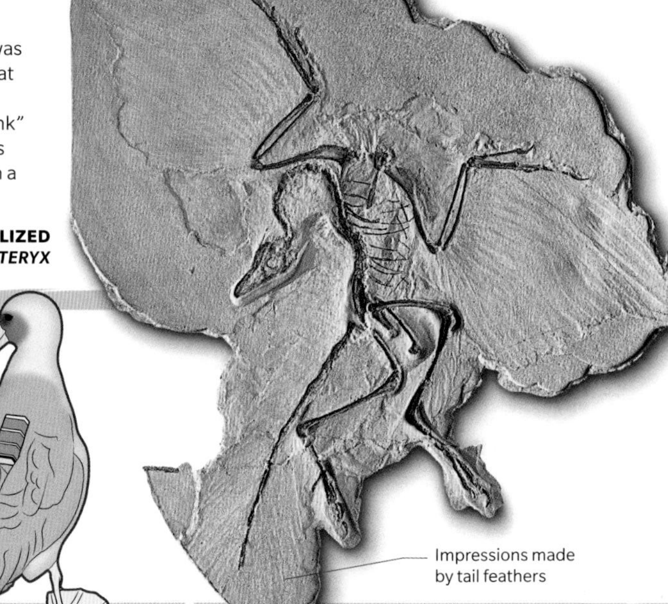

FOSSILIZED *ARCHAEOPTERYX*

1861
A fossilized *Archaeopteryx* was found in Jurassic limestone at Solnhofen in Germany. The discovery of this "missing link" in the evolution of dinosaurs into birds was announced in a paper published in 1861.

1889
The Society for the Protection of Birds was founded by English philanthropist Emily Williamson in the UK. It campaigned against decorating hats with feathers from birds such as little egrets, thus driving them toward extinction.

EMILY WILLIAMSON

1989
Albatrosses were the first birds to be tracked remotely by satellite. This revolutionized scientists' understanding of the vast journeys undertaken by these birds when feeding at sea, and the movements and migratory habits of many other species.

ALBATROSS WITH TRACKER

One of three clawed digits on each wing

ARCHAEOPTERYX

> "Birds without exception lay eggs, but the pairing season and the times of parturition are not alike for all."
>
> ARISTOTLE, *The History of Animals, Book VI*, 4th century BCE

c.160–150 MYA
Birds evolved from carnivorous dinosaurs in the Late Jurassic. Like modern birds, *Archaeopteryx* had long wing and tail feathers, but it also had teeth. It could fly weakly.

c.127–121 MYA
About the size of a hummingbird, *Liaoxiornis* was a primitive bird that lived in eastern Asia. It had more powerful flight muscles than earlier birds, and its feet could grasp a branch.

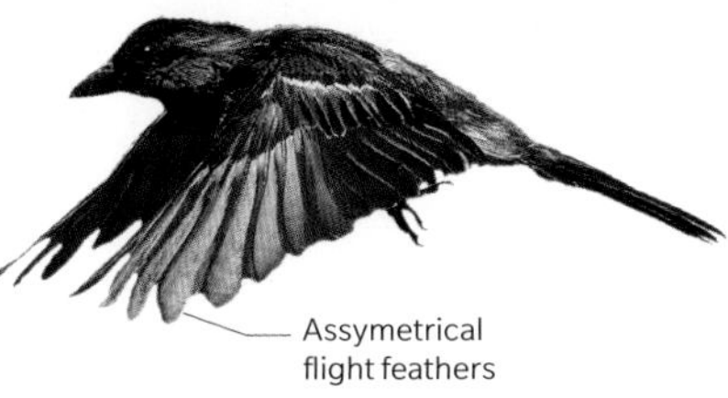

LIAOXIORNIS

FIRST PASSERINE

c.52 MYA
The first-known passerine, or bird with perching feet, appeared. Called *Eofringillirostrum boudreauxi*, it is the earliest bird known to have a finchlike beak.

c.66 MYA
When an asteroid hit Earth, all nonavian dinosaurs and bird groups other than the toothless beaked birds were wiped out. These birds had a powerful gizzard capable of crushing tough seeds.

c.66.7 MYA
The first-known modern bird, *Asteriornis maastrichtensis*, lived alongside its dinosaur relatives. Nicknamed the "wonderchicken," it had a face like a modern chicken, including a toothless beak, but the back of its skull was like that of duck.

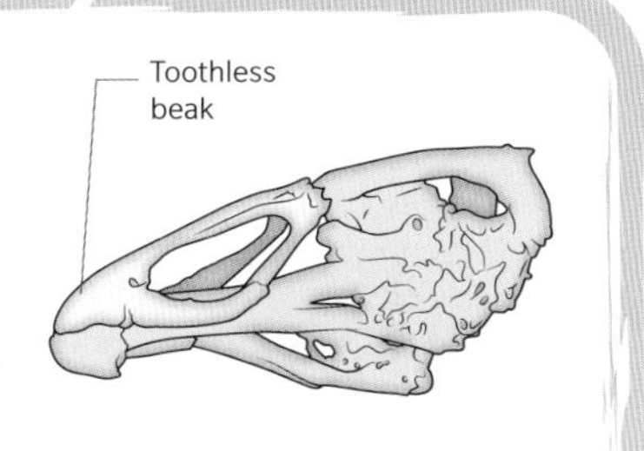

WONDERCHICKEN SKULL

c.1485–1506
Italian polymath Leonardo da Vinci made detailed studies and sketches of the flight of birds and various flying machines, including some designs for a mechanically powered ornithopter. In 1505–06, he produced the *Codice sul volo degli uccelli* (Codex on the flight of birds).

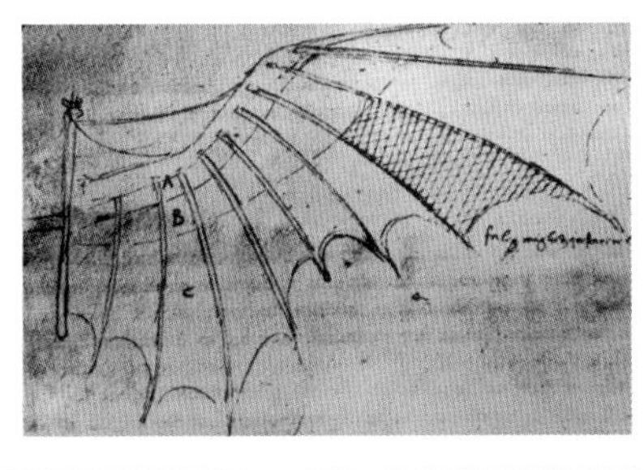

ORNITHOPTER SKETCH

c.16th century
The last of the giant bird species, the flightless moas, were hunted to extinction by Māori in New Zealand. Standing at $11\frac{3}{4}$ ft (3.6 m), *Dinornis* was the tallest bird that ever lived.

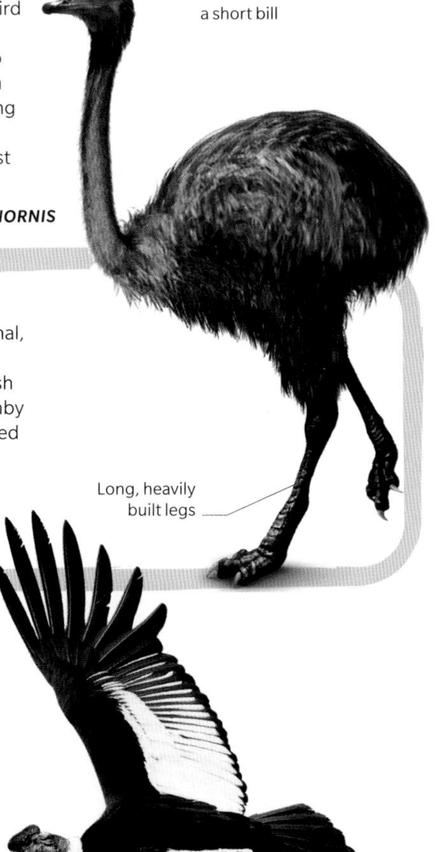

DINORNIS

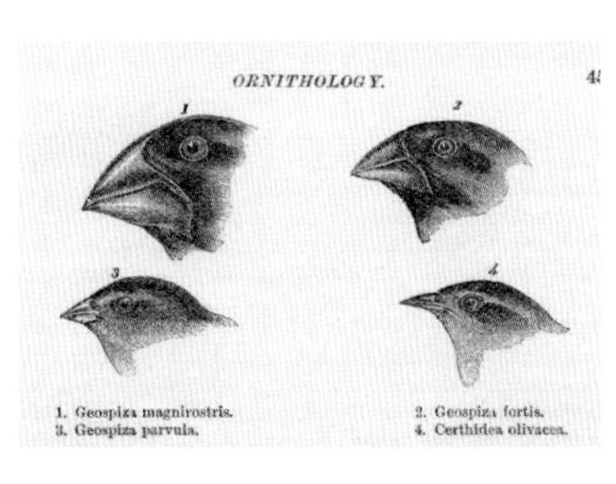

DARWIN'S FINCHES

1835
British naturalist Charles Darwin's voyage on HMS *Beagle* reached the Galápagos Islands, where he collected species of finches with markedly different bills.

1676
Ornithology, the first rational, scientific classification of birds, developed by English naturalists Francis Willughby and John Ray, was published in Latin. An enlarged and corrected English version was published in 1678.

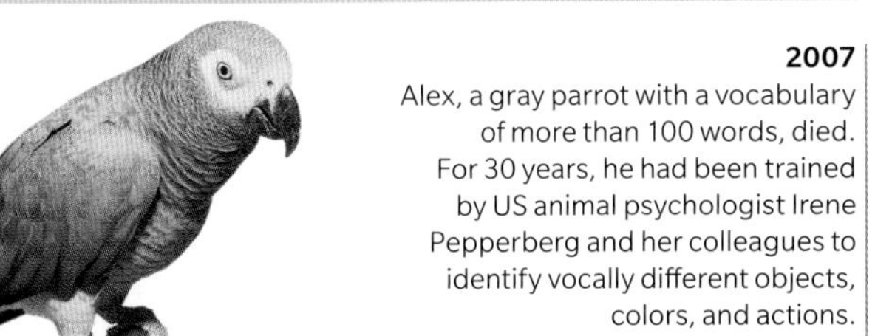

GRAY PARROT

2007
Alex, a gray parrot with a vocabulary of more than 100 words, died. For 30 years, he had been trained by US animal psychologist Irene Pepperberg and her colleagues to identify vocally different objects, colors, and actions.

2023
Scientists discovered a huge pile of guano around an Andean condor nest in northern Patagonia, Argentina. Generations of these vultures had been adding to it for 2,200 years.

ANDEAN CONDOR

Feathers

Insulation, waterproofing, color, shape—feathers perform many functions as well as enabling birds to fly. Their individual shapes combine to create wings for many purposes, from fast flight to slow soaring, and from short flits to long migrations.

A **swan** may have more than **25,000 feathers,** while a **hummingbird** may have fewer than **1,000**

Feather colors

Colors are vital to many aspects of birds' lives. Bright colors can be used to impress a rival or prospective mate, while cryptic patterns can help a bird hide. Most feather colors are created by pigments, some from food, others generated internally. Other colors are the result of microscopic structures in the feathers scattering incoming light, with some wavelengths being absorbed and others reflected. Shining iridescence occurs when two close surfaces reflect light—as the viewing angle changes, different colors become visible.

GALAH
Eolophus roseicapilla

NORTHERN CARDINAL
Cardinalis cardinalis

AMERICAN ROBIN
Turdus migratorius

BLACK-NAPED ORIOLE
Oriolus chinensis

Caring for feathers

Feathers need to be constantly cleaned and cared for to keep them in good condition. The preen (uropygial) gland above a bird's tail secretes oils that are transferred to the feathers by preening. The oils may help waterproof and maintain feather structures, deter parasites, and even create a scent, depending on the species.

➤ **Preening for perfection**
This Bald Eagle (*Haliaeetus leucocephalus*) is "zipping" one of its tail feathers back into perfect shape by drawing the feather through its bill.

TYPES OF FEATHER

The barbs of a feather are loose or tightly connected into flat surfaces (vanes) beside a stiff shaft. The simplest feather is a sensory filament, called a filoplume. The most complex feathers have a tough, rigid shaft with a hollow base and a vane each side, with a constant shape according to location.

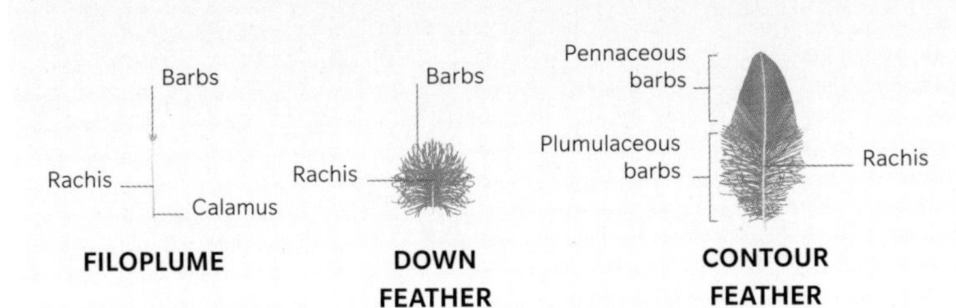

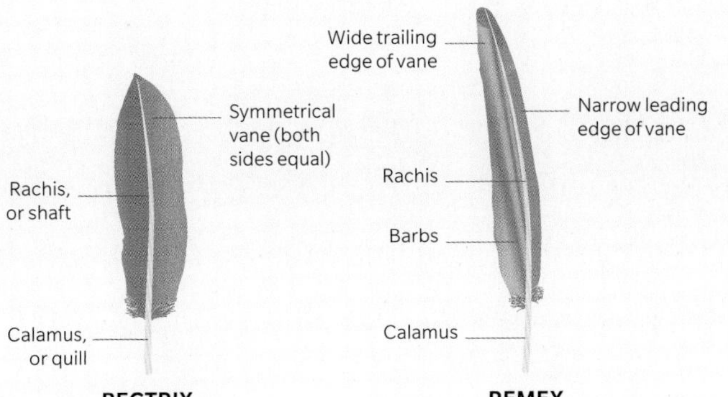

VERNAL HANGING PARROT
Loriculus vernalis

BLUE DACNIS
Dacnis cayana

BLUE JAY
Cyanocitta cristata

VIOLET-BACKED STARLING
Cinnyricinclus leucogaster

Several pigments combine with iridescence to create shining purples

▲ Band of colors
The world of birds has a complete spectrum of rich, vibrant color and a huge variety of patterns, mostly created by their feathers.

Flight feathers structure

Wing flight feathers must press hard against air during powerful downbeats, and twist to allow air to pass on the upbeat. A narrow outer vane and a softer, broader inner vane help achieve this. Each vane consists of a sheet of parallel barbs that branch into tiny interlocking barbules, creating a smooth, aerodynamic surface. The tail flight feathers are used for braking and steering.

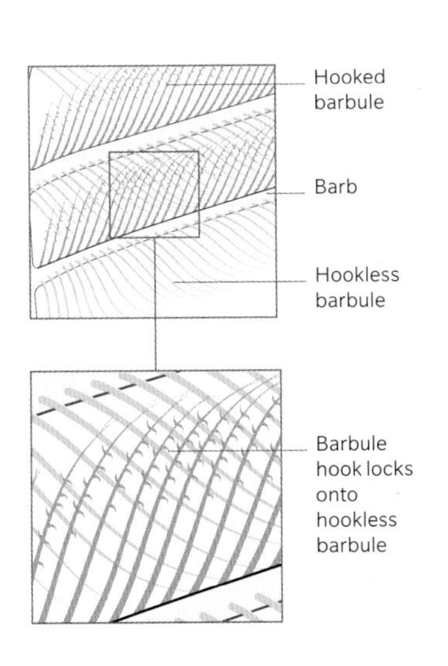

► Life in black and white
Plumage looks white when all wavelengths of light are reflected by the structure of the feathers, whereas black feathers are the result of melanin pigment granules made in skin cells.

ADELIE PENGUIN
Pygoscelis adeliae

Flight

A large part of the appeal of birds to humans comes from their ability to fly. Our sense of wonder and envy is only increased by a basic knowledge of how different birds are able to achieve such amazing feats in the air.

When hovering, a **hummingbird's heart** may **beat** at a rate of **1,200 times** per minute

Slow flapping
Many birds, including gulls, flap their wings slowly in direct flight, often between short glides.

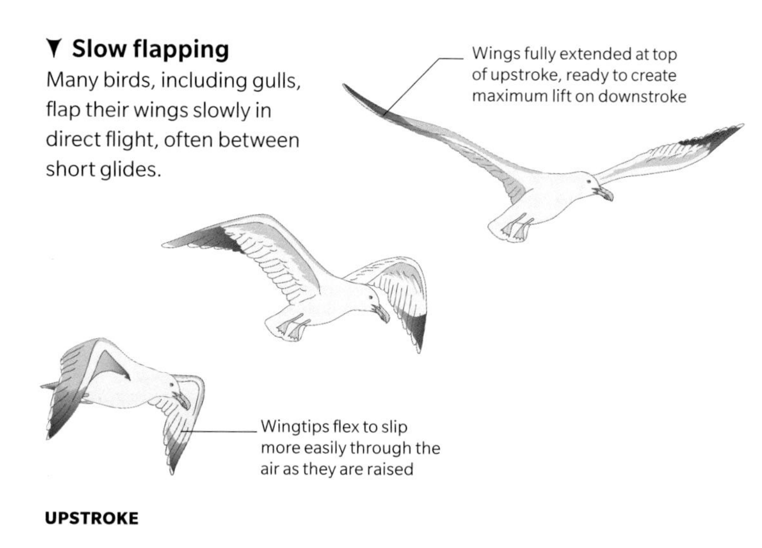

Wings in position for next upstroke

Flapping flight

To achieve flapping flight, birds use their powerful breast and wing muscles to move their wings in a repetitive motion. At takeoff, downward wingbeats compress the air underneath the wings, pushing the bird upward. As the flapping speed increases, air rushing over the wings generates lift, allowing the bird to maintain altitude, while each successive flap creates thrust to move it forward.

Gliding and soaring

Gliding is forward flight on spread wings without beats. A longer convex upper surface forces air over the wing slightly faster than that beneath the flat underside, creating a pressure difference that generates lift. Larger birds soar by circling in columns of warm, rising air, called thermals, that carry them up. Smaller birds lack the necessary wing surface area to glide or soar. However, they can fly rapidly by flapping their wings in short bursts then holding them against their sides, so they shoot forward using momentum, rather than lift.

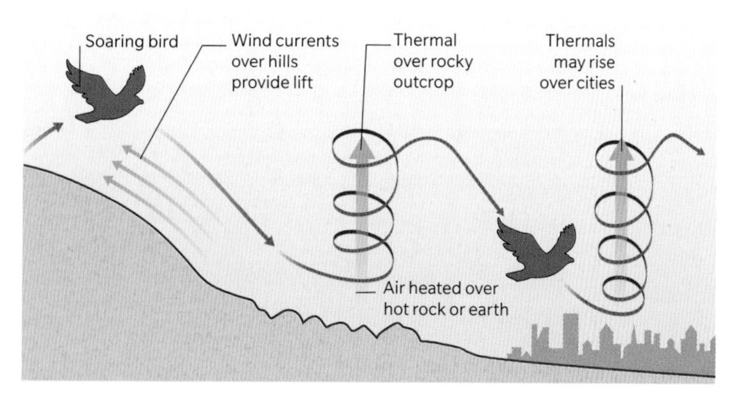

Thermals and updrafts
Larger birds, such as birds of prey, gulls, storks, and pelicans can rise and stay aloft on thermals without having to expend any energy. Updrafts caused by deflected wind currents can also provide enough lift for soaring.

Seabird turns or swoops down, gaining speed using gravity

Wind direction

Seabird rises, banking steeply against strong wind

Dynamic soaring
Ocean winds are slowest near the surface due to drag and faster higher up. Seabirds allow the high, fast winds to carry them before dropping down to the surface, where they turn into the wind and are lifted up again.

➤ Holding steady
Its rapid wingbeats enable the Green-breasted Mango (*Anthracothorax prevostii*) to hold its position in front of nectar-rich flowers as it feeds.

Hovering

Hovering entails holding steady, keeping a stable position relative to the ground. To achieve this, birds such as terns, kingfishers, and some birds of prey adjust their wingbeats so that their flight speed matches that of the wind. Facing into the wind, they make precise adjustments in order to hold position, their tail helping with stability. Hummingbirds take hovering to a different level, with special adaptations enabling an extraordinary physical performance.

➤ How a hummingbird is able to hover
A special joint and unique humerus allow hummingbirds' extremely rapid wingbeats. Very strong muscles attached to the breastbone and a faster metabolic rate than any other animal help take their flight to an unsurpassed level.

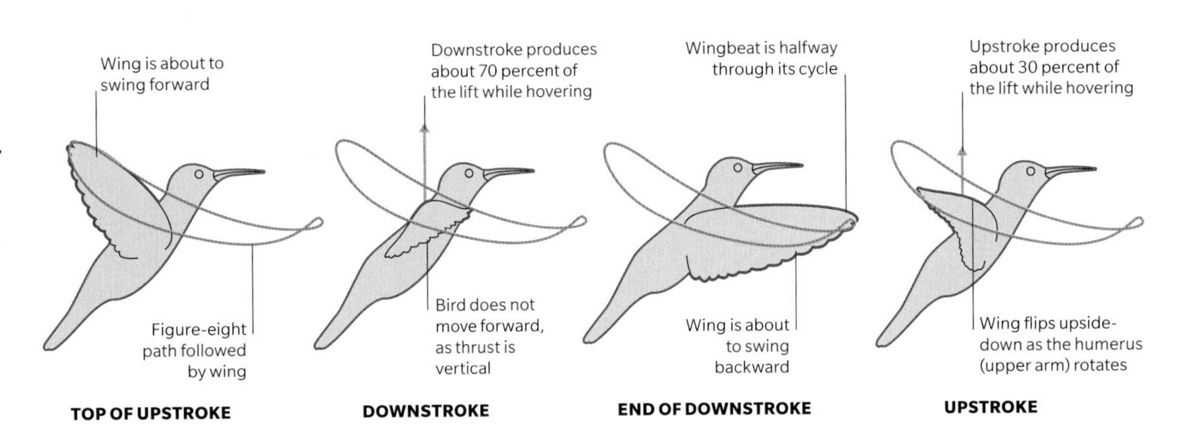

Flying underwater

Auks and penguins can "fly" underwater. Many other species, such as seaducks and gannets, also use their wings when submerged. Much the same dynamics apply as in the air. A narrow surface cuts through water on the forward or upstroke, then the wing is turned against the water for maximum power on the backstroke.

▼ Streamlined body
The torpedolike form and stiff, flipperlike wings of a King Penguin (*Aptenodytes patagonicus*) are ideal for propulsion underwater. It can dive as deep as 1,000 ft (300 m).

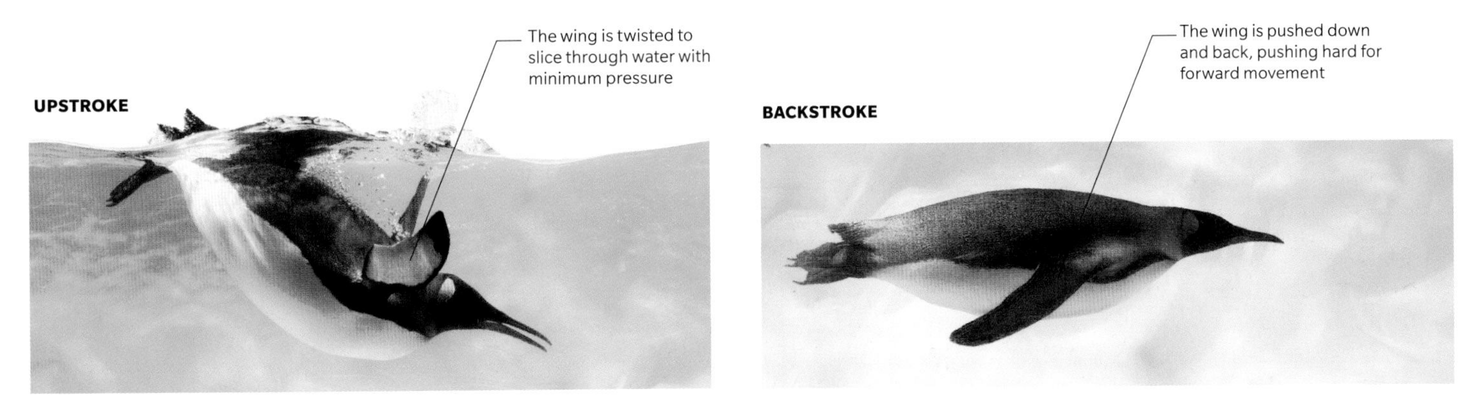

The gentoo penguin can swim almost as fast underwater as the 100 m world record holder, Usain Bolt, can sprint on land

Senses

Bird senses parallel our own but, astonishingly for their size, the sensitivity of some extends well beyond that of humans. Many species have exceptional color vision and can also see ultraviolet light, while others have a remarkable sense of hearing or smell. And some birds use the Earth's magnetic field to help them navigate.

Sight

Birds of prey, such as owls, have eyes that face forward, giving them a large field of binocular vision, which allows them to judge distances very accurately. Other species have eyes on either side of their head, and this monocular vision allows them to focus each eye on different objects simultaneously, helping them avoid predators. Day-active birds have excellent color vision and they can also see ultraviolet light, both of which help them find food and choose a mate.

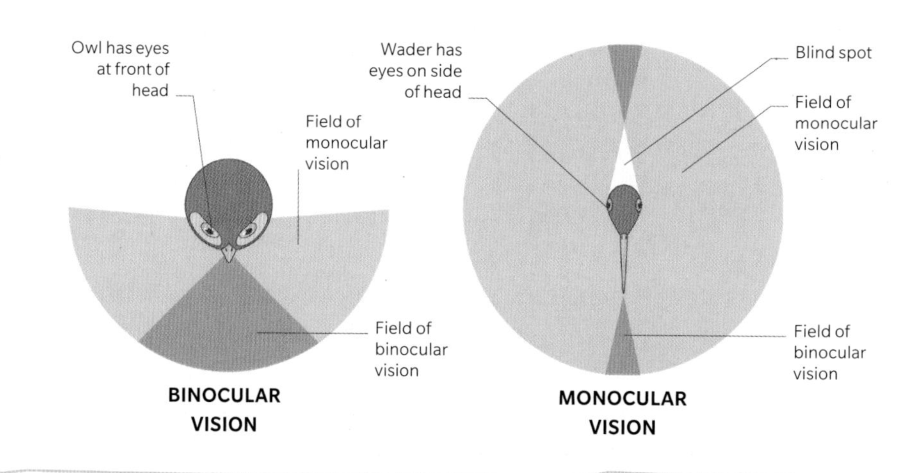

Hearing

Birds lack external ears and the openings are hidden below their plumage. In general, their hearing is not much more acute than our own, but migratory species may be able to hear very low-level sound, perhaps waves on distant shores. Night-hunting owls use their sensitive hearing to locate their prey, and can even hear small rodents moving in tunnels under snow.

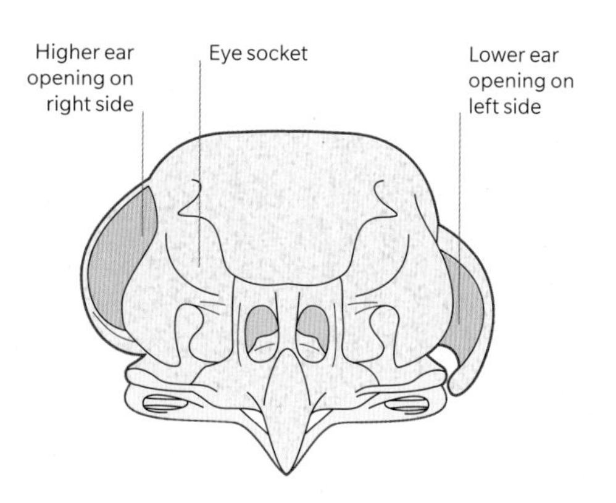

▲ Assymetrical ears
This Boreal Owl (*Aegolius funereus*) skull illustrates a typical owl feature, where one ear opening is higher than the other, enabling them to pinpoint the source of a sound.

Migration

Many species migrate as a family or in flocks, while other species go alone with no experience of timing, direction, or routes. At the right time, if conditions are good, they obey the urgings of their internal clock to depart, often at night. Birds use a variety of sensory cues to find their way to their destination—a breeding site or feeding grounds—and they have a special light-sensitive protein in their eyes that helps them detect the position and strength of the Earth's magnetic field. Some species can navigate distances of thousands of miles.

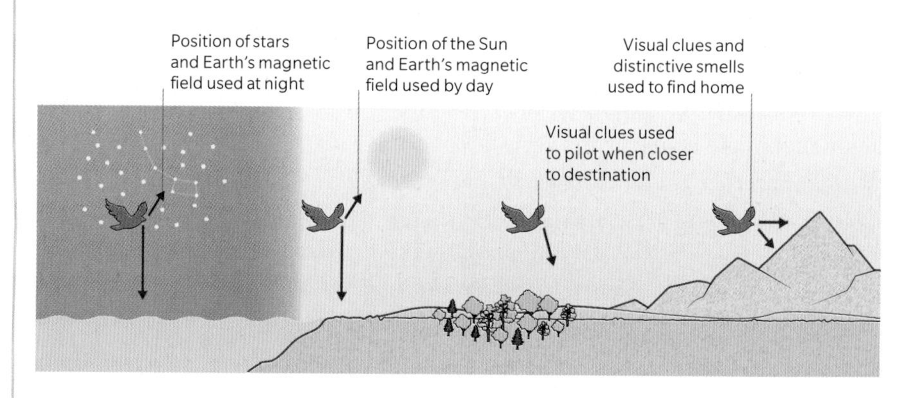

▲ How birds navigate
On initial migratory flights, birds must rely on clues such as positions of stars and the Sun, as well as the Earth's magnetic field.

Repetition brings experience, so birds can then use landmarks: coastal features, mountains and rivers, and smaller details when close to home.

Visual displays

Courtship and territorial defense rely on visual displays, including ritualized movements, that exploit the sensitive eyesight and appreciation of colors and patterns possessed by many species. Remarkable ornamentation and coloration have evolved to prove the fitness of the male and to impress with his beauty. And while both sexes of many songbird species look the same to human eyes, most of the males have areas of plumage that reflect ultraviolet light. Ornamentation may be in the form of elongated or specially shaped feathers, especially on the head and tail—the peacock's train is a famous example—or bizarre growths on the head or bill. Extreme features can sometimes impair the male's physical abilities, such as flight, making it a fine balance between individual survival and successful reproduction.

▲ Sensational shimmer
A shimmer of color and pattern on a huge, widely spread fan characterizes the male Indian Peafowl (*Pavo cristatus*) as it displays its remarkable eyespots.

Vocal communication

Birds communicate important messages vocally. Frequent simple contact calls keep flocks and families together as they forage. Sharp, agitated notes warn others, even different species, that a predator is near. Wading birds make loud, repeated calls as they take flight, and songbirds keep in contact with various trills and chatters as they fly. Song, used to defend a territory or attract a mate, may be a repeated phrase or more improvised, revealing experience and quality. While it is usually the males that sing, many females also do so intermittently.

Air expelled through the mouth vibrates flexible membranes of the syrinx (voice box), a structure at the base of the trachea (windpipe)

► Territorial song
Songbirds, such as this Bluethroat (*Luscinia svecica*), have a built-in song that they may develop through copying and mimicry.

Feeding

Many aspects of their habitat influence the distribution of bird species. However, it is their basic need for a constant supply of suitable, nutritious food to support their high-energy lifestyles that determines where they will be at any given time.

Scientists **estimate** that birds may eat up to **500 million tons** of insects **every year**

▲ Seeds
A European Goldfinch (*Carduelis carduelis*) is able to reach between the spines of a teasel seed head with its slender, pointed bill in order to extract the highly nutritious seeds without difficulty.

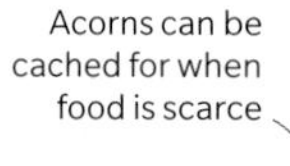

▲ Nuts
Woodpeckers can open tough nuts by wedging them in crevices. A Eurasian Jay (*Garrulus glandarius*) will bury hundreds of acorns and can find them weeks later, eating them when other food is scarce.

▲ Fruit
Bohemian Waxwings (*Bombycilla garrulus*) are insectivorous in summer but need a plentiful supply of berries in winter. In some years, they may travel thousands of miles to find a good crop.

Eating plants

Some birds eat leaves, but many more species feed on nectar, seeds, and fruit. Grazers, such as geese, pluck leaves and roots as they walk over the ground, while swans dip for aquatic plants, and swamphens rip open large stems for the nutritious center. More birds eat seeds, which are available year-round, than any other plant-based food. Berries and nuts offer a more seasonal food.

► Drinking nectar
Asia's Streaked Spiderhunter (*Arachnothera magna*) probes into a blossom to lap up nectar, just like New World hummingbirds and Old World sunbirds and honeyeaters. Shorter-billed species pierce blooms at the base.

Digestion

An elastic crop briefly stores food before it is digested, allowing a bird to gather large quantities while it can. A muscular gizzard then grinds it down. Strong stomach acids help vultures eat rotten meat, and even bones may be digested within 24 hours. At the other extreme, tiny nectar-feeders have a very fast metabolism and can digest most food within an hour.

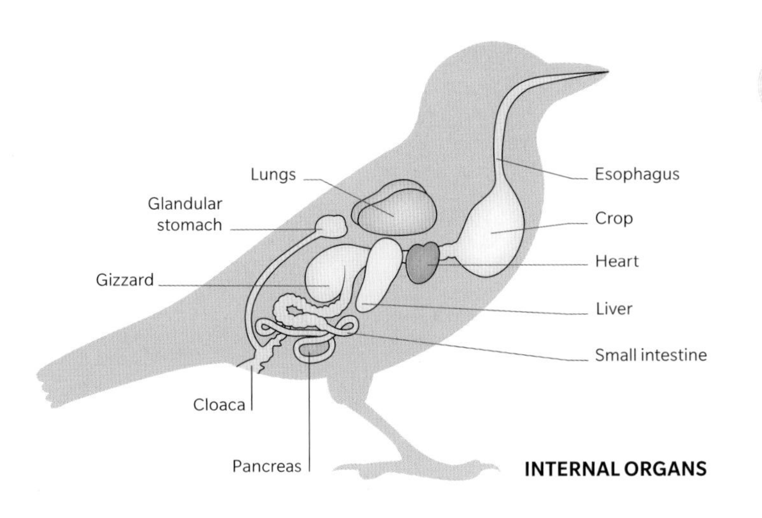

INTERNAL ORGANS

A beakful of insects suggests hungry chicks are nearby

▲ Insects

Insects are vital to the whole food chain, and millions of birds rely on them. The Gray Wagtail (*Motacilla cinerea*) forages alongside water, especially fast-running, clean streams and rivers.

Once caught, fish are typically swallowed whole

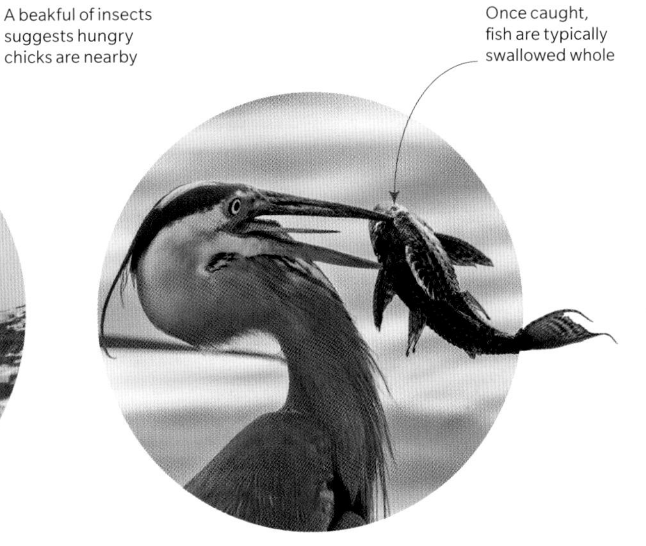

▲ Fish

Many seabirds eat fish. Some freshwater species catch fish underwater, others by waiting patiently at the water's edge, ready to strike. The Great Blue Heron (*Ardea herodias*) grabs fish with its bill.

Small prey is swallowed whole, but larger prey is first torn apart

▲ Meat

Like other barn owls, the Eastern Barn Owl (*Tyto javanica*) detects live prey as often by sound as by sight. Indigestible parts of its meal, such as fur and bones, will be regurgitated in pellet form.

Eating animals

A vast array of bird species eat live food, from worms and larvae to fish, rodents, and other birds. Shorebirds probe mud for marine worms and pry shellfish from rocks. Warblers meticulously glean insects from foliage. Birds of prey include specialists that catch birds, fish, and large insects. Kites, vultures, condors, and crows also take a great deal of carrion (meat from animals found already dead).

► Using tools

Very few birds use tools. New Caledonian Crows (*Corvus moneduloides*) use sticks to pull larvae from holes, while Egyptian Vultures (*Neophron percnopterus*) toss stones at ostrich eggs to break them open.

Inspired by Birds

For millennia, humans have been fascinated with birds—they feature in the mythology and religions of many different cultures, and have become symbols of authority and grace. They have been traded for their beautiful plumage, valued as songbirds, and loved as pets. And birds have inspired countless artists, poets, authors, and even film directors.

500 BCE–500 CE
The Nasca people covered a desert near the coast of Peru in hundreds of geoglyphs. Some of the bird outlines created include a pelican, a baby parrot, and a bird of prey that used to be identified as a condor, but which experts now think could possibly be of an extinct species.

VULTURE GEOGLYPH

6th century
Quill pens replaced hollow reeds as the principal instrument used for writing in medieval Europe. The best feathers to use were the first five primaries, most commonly obtained from a goose.

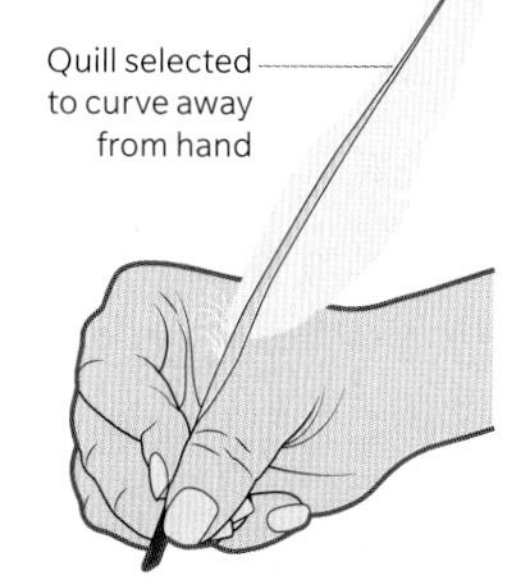

QUILL PEN

ILLUSTRATED MANUSCRIPT

c.1177
Persian poet Farid ud-Din Attar completed the *"Mantiq al-tayr"* ("Conference of the Birds"), an allegory of the human soul's search for meaning, in which 30 birds go on a quest to find their supreme master, the Simurgh, the legendary bird of Iranian folklore.

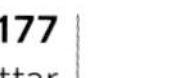

FREDERIK II

c.1250
Holy Roman Emperor Frederik II writes *De Arte Venandi cum Avibus* (*The Art of Hunting with Birds*), a history of falconry, and the first illustrated ornithological text.

1885
The jeweler Carl Fabergé created the Hen Egg in St. Petersburg, Russia. This was the first imperial Easter egg, which was given by Tzar Alexander III to his wife, Maria.

FABERGÉ EGG

1934
Donald Duck, the irascible cartoon character created by the Walt Disney Company, made his first theatrical appearance in the US.

c.1947
As air travel became more common after WWII, some airlines started to use posters featuring native birds, such as this black grouse, to entice adventurous tourists to book a trip.

NORWEGIAN TRAVEL POSTER

1952
Colonel Harland Sanders opened the first Kentucky Fried Chicken (KFC) franchise near Salt Lake City, Utah, US. Today, there are more than 25,000 KFC franchises worldwide, consuming nearly 1 billion chickens a year.

THE BIRDS

1963
British director Alfred Hitchcock's film *The Birds* was released. Its depiction of the inhabitants of a small California town being terrorized by a series of unexplained vicious attacks by birds led to a rise in ornithophobia (fear of birds).

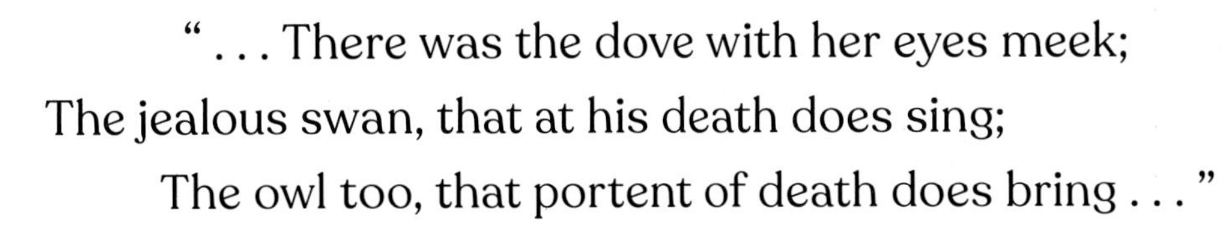

"... There was the dove with her eyes meek;
The jealous swan, that at his death does sing;
The owl too, that portent of death does bring ..."

GEOFFREY CHAUCER, "Parlement of Foules," c.1380

NEOLITHIC WALL ART

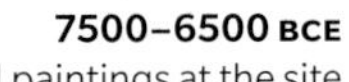

7500–6500 BCE
Wall paintings at the site of Çatalhöyük in Turkey, where the deceased were placed under the floors of houses, show griffon vultures scavenging headless bodies.

7th century BCE
A young woman, who is assumed to have died in childbirth, was buried with her son, who was laid on a swan's wing. Their grave is at Vedbaek, eastern Denmark.

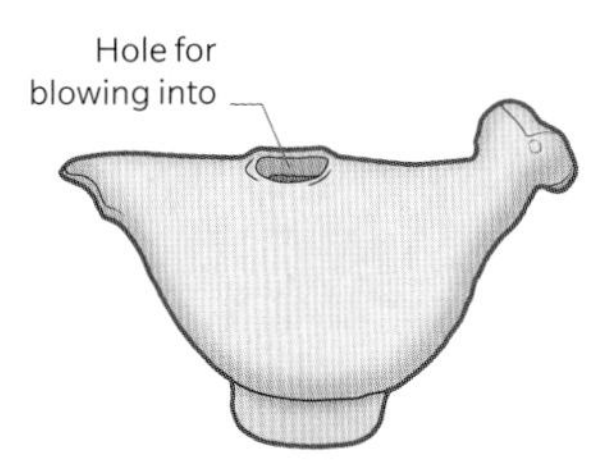

BIRD WHISTLE

3000 BCE
The Indus Valley and Helmand civilizations made simple clay and terracotta bird-shaped whistles as musical toys for children.

1600–1027 BCE
Birds held significant roles in Chinese mythology during the Shang Dynasty, and many vessels were cast in the form of an owl. As a symbol of death, owl vessels may have been used in rituals for the dead.

OWL-SHAPED BRONZE VESSEL

2686–30 BCE
Some ancient Egyptian gods were depicted as birds. Originally worshipped as a sky god, as indicated by his falcon form, Horus came to embody divine kingship and pharaohs ruled as his living representative.

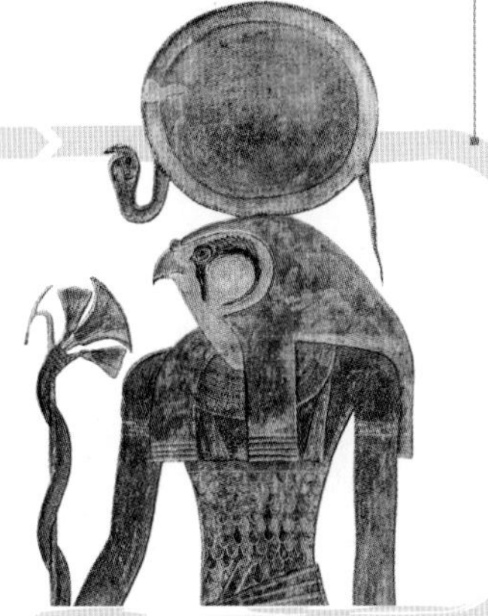

HORUS

13th–19th centuries
In the kingdom of Benin, cockerels were a symbol of strong leadership. Bronze casts were placed on palace altars in tribute.

BRONZE COCKEREL

c.1380
English poet Geoffrey Chaucer wrote the dream-vision poem "Parlement of Foules" in which a group of birds gather on "seynt valentynes day" in early spring to choose their mates for the upcoming year.

17th century
In Europe, decorative oil paintings became fashionable. Whimsical paintings of different species of birds perched in a tree with a musical score were popularized by the Flemish artist Frans Snyders.

CONCERT OF THE BIRDS

1827–1838
French-US artist and naturalist John James Audubon's *Birds of America* was published as a series of life-size prints. All 435 bird species were depicted in their natural habitat.

AMERICAN FLAMINGO

c.1800s
Money coils made with the scarlet feathers of the cardinal myzomela were traded in the Santa Cruz Islands. Known as *tevau*, their value as currency was determined by the size and quality of the feathers.

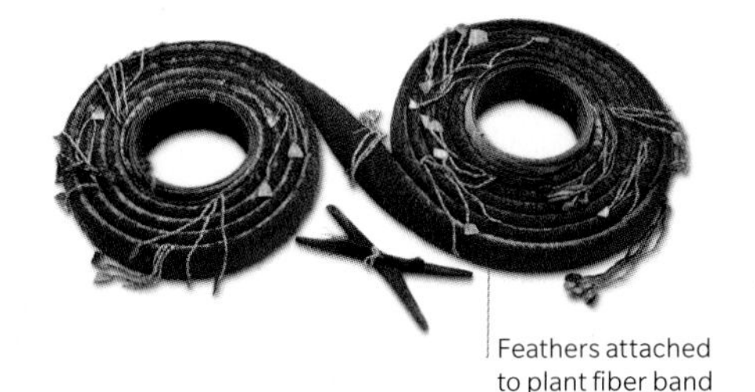

FEATHER CURRENCY

2016
Arjan Dwarshuis of the Netherlands observed a record 6,852 bird species in the field in one calendar year. In 2024, American Peter Kaestner set a new lifetime record for the most species observed: 10,000.

BIRDWATCHING

2020
A new world record for the most expensive bird was set on November 15, 2020, when New Kim, a two-year-old champion racing pigeon, was sold at auction in Belgium for $1.8m (£1.4m).

RACING PIGEON

Breeding

While the first priority of any bird is simply survival, its next goal must be to reproduce in order to continue the line and guarantee the future security of its species. Rearing this next generation can be intense work for the parents.

Nests

A nest is a place to lay eggs and, often, to rear chicks until they can fly. It is not in any other sense a "house," although some may be used for roosting outside of the breeding season. Nests vary from a bare ledge to massive social structures (as with the Sociable Weaver, *Philetairus socius*). Nearly all bird species have a single nest, from a scrape on the ground to complex, beautifully built structures of twigs, moss, feathers, and even cobwebs. Natural cavities may be used, or other species may take over holes excavated by woodpeckers or old stick nests built by crows. Depending on the type of structure, a nest can insulate eggs and chicks and conceal them from predators.

▲ Gathering nest material
Each species follows strict requirements for nesting sites and materials. This Blue Tit (*Cyanistes caeruleus*) will nest inside some kind of cavity, using moss with a lining of whatever feathers and fur it can find.

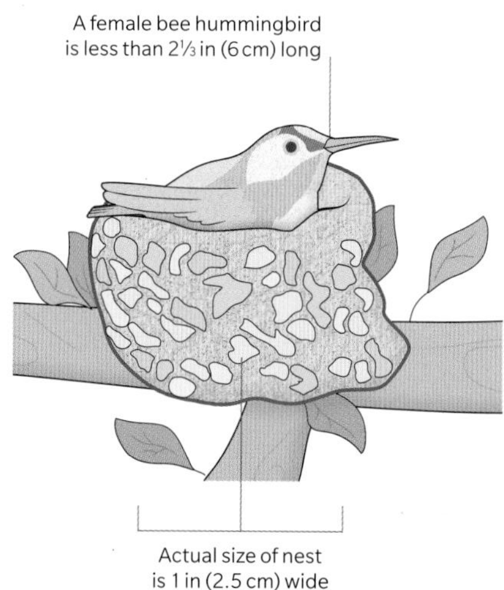

▲ Cup nest
Cup-shaped nests are generally well-hidden, but the Bee Hummingbird (*Mellisuga helenae*), which makes the smallest of all at 1 in (2.5 cm) across, simply mimics a bulge on an open branch.

◄ Platform nest
Many species, from herons and buzzards to crows, build large stick nests. The White Stork (*Ciconia ciconia*) creates a huge structure, which a pair may use for several years, on a tree or building.

Eggs

Birds' eggs have hard but porous calcium shells, which may be round, oval, or pointed at one end. Colors and patterns develop just hours before an egg is laid. With the exception of the Malleefowl (*Leipoa ocellata*), which incubates them in the heat of decaying vegetation, birds incubate eggs by sitting on them (usually against a "brood patch" of hot, bare skin), until they hatch. Some species produce a single egg, while others may lay more than 15.

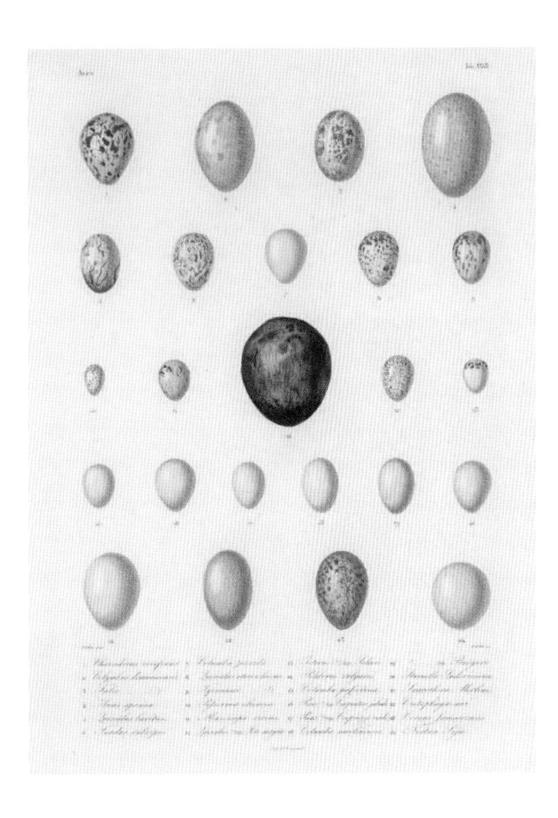

➤ **Colors and patterns**
A selection of eggs from Cuba shows a typical variety of color. White eggs are visible in dark cavities; those in more open nests are often camouflaged.

CONSERVATION

Most species nest successfully given suitable habitat and food. Others require active conservation measures, such as terns and plovers nesting on sand or shingle beaches, and egrets in waterside trees (requiring special protection of colonies). Species such as the eastern bluebird (*Sialia sialis*) in the US benefit from nestbox schemes.

EASTERN BLUEBIRD

Parental care

Eggs hatch after an incubation of around 12–14 days (the smallest species) to 80 days (larger albatrosses). This may be by a single parent (usually female, but in some species the roles are reversed) or by both. Chicks are fed in the nest until they fly (fledge) or are cared for nearby as they learn to find food themselves. Typical small songbirds are fed for two or three weeks until they fly and for a similar period afterward. A few species have "helpers" (often juveniles from earlier broods) feeding chicks as well as the parents. Brood parasites, such as the Common Cuckoo (*Cuculus canorus*), take no part in rearing their own chicks, instead laying eggs in the nest of another species and relying on the hosts to function as foster-parents to the interloper.

▲ **Altricial**
Species with enclosed or cup nests, such as this European Robin (*Erithacus rubecula*), tend to have chicks that hatch blind and naked and remain in the nest while their first feathers grow.

▲ **Semi-altricial**
Penguin chicks living in extreme conditions, such as Emperor Penguins (*Aptenodytes forsteri*), hatch with a warm, downy covering, but must still be brooded and fed by their parents until they have a full covering of feathers and become independent.

▲ **Precocial**
Ground-nesters and Hooded Mergansers (*Lophodytes cucullatus*), which nest in tree cavities, may hatch with warm down and eyes open. They often feed themselves as their feathers develop.

Classifying Birds

Classification is the science of identifying and categorizing living things, based on physical features and other inherited characteristics, including DNA. There is still much debate among ornithologists as to which classification groups different birds belong.

Classification groups

Scientists group living things into a series of levels. The highest, most inclusive levels are kingdoms, such as the animal or plant kingdoms. These are subdivided into phyla, then classes. Each class is subdivided in turn, to describe every species (see panel, right). A species is a group of similar individuals that are able to interbreed in the wild and cannot breed successfully with other species (unlike domesticated "breeds" of the same species). Some species have distinct populations that vary from each other in signficant ways, and these forms are known as subspecies (or races).

Passerines

The order Passeriformes contains more than half of the world's bird species. Commonly known as perching birds, they have three forward-facing toes and one hind toe. There are two suborders, one with more primitive species, the other a much larger group, the songbirds. Passerines are found in almost every terrestrial habitat, from mountains to coasts, but not at sea. Boasting a huge array of shapes, sizes, and colors, passerines include many familiar birds, such as sparrows and swallows, as well as localized families, such as New Guinea's birds-of-paradise.

▼ Lockable tendons

Passerines grip a perch automatically. Simply resting their weight upon it stretches tendons, causing the toes to curl inward in a tightening grip, which is "locked" in position.

Front flexor tendon

Rear flexor tendon

Ribbed edge of tendon

Ribbed surface of tendon sheath

STANDING ON A FLAT SURFACE

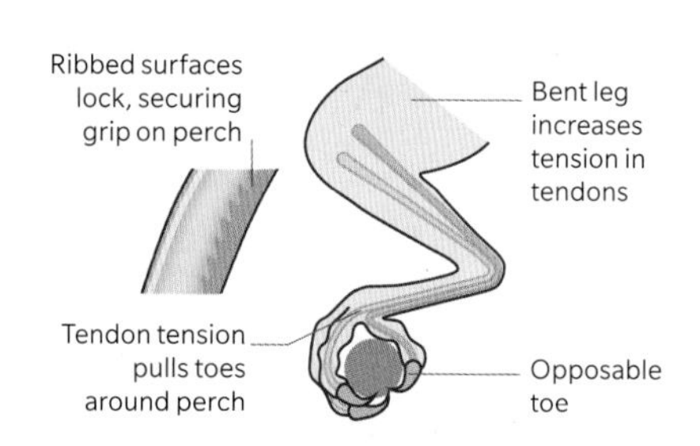

GRIPPING A PERCH

HOW CLASSIFICATION WORKS

The major classification levels are listed here, using the common starling as an example. Below species level, distinct variants are classed as subspecies.

▼ CLASS Aves
Aves contains all of the world's birds. They are the only group of animals alive today that have feathers. Most, but not all, birds can fly.

▼ ORDER Passeriformes
The largest order of birds, containing more than 6,500 species. All members have specialized feet for gripping perches. Many have complex songs.

▼ FAMILY Sturnidae
This family contains all of the world's starlings, more than 120 species. Starlings have a straight bill and an omnivorous diet, and they walk rather than hop.

▼ GENUS *Sturnus*
This genus contains two species of starlings. They are distinguished by differences in their plumage and geographic distribution as well as their DNA.

▼ SPECIES *Sturnus vulgaris*
The common starling has dark, iridescent plumage with pale spots, which wear off by the breeding season. It often roosts in large flocks.

▲ Changing classification
The *Edinburgh Journal of Natural History* (1835–40) shows several hummingbirds in the genus *Trochilus*, which today contains just two of the world's 366 species.

BIRD ORDERS

The 44 orders listed below follow the names and classification sequence of the World Bird List released by the International Ornithological Committee (IOC) in 2023. In this sequence, more than 11,000 bird species are recognized, belonging to more than 250 families grouped into the orders. New research regularly recommends elevating certain subspecies to species status or combining previously separated subspecies into one, or rearranging families.

➤ Flightless birds

At various points in evolution, different groups of birds lost the ability to fly. Today, the land-dwelling kiwis, for example, have tiny vestigial wings, whereas penguins evolved paddlelike wings with which they "fly" underwater.

NORTH ISLAND BROWN KIWI
Apteryx mantelli

COMMON NAME	ORDER	FAMILY
Ostriches	Struthioniformes	1
Rheas	Rheiformes	1
Kiwis	Apterygiformes	1
Emu and Cassowaries	Casuariiformes	1
Tinamous	Tinamiformes	1
Waterfowl	Anseriformes	3
Gamebirds	Galliformes	5
Nightjars	Caprimulgiformes	1
Oilbird	Steatornithiformes	1
Potoos	Nyctibiiformes	1
Frogmouths	Podargiformes	1
Owlet Nightjars	Aegotheliformes	1
Swifts and Hummingbirds	Apodiformes	3
Turacos	Musophagiformes	1
Bustards	Otidiformes	1
Cuckoos	Cuculiformes	1
Mesites	Mesitornithiformes	1
Sandgrouse	Pterocliformes	1
Pigeons and Doves	Columbiformes	1
Cranes and relatives	Gruiformes	6
Grebes	Podicipediformes	1
Flamingos	Phoenicopteriformes	1
Waders, Gulls, and Auks	Charadriiformes	19
Kagu and Sunbittern	Eurypygiformes	2
Tropicbirds	Phaethontiformes	1
Loons	Gaviiformes	1
Penguins	Sphenisciformes	1
Albatrosses and Petrels	Procellariiformes	4
Storks	Ciconiiformes	1
Cormorants and relatives	Suliformes	4
Pelicans and relatives	Pelecaniformes	5
Hoatzin	Opisthocomiformes	1
Birds of Prey	Accipitriformes	4
Owls	Strigiformes	2
Mousebirds	Coliiformes	1
Cuckoo-roller	Leptosomiformes	1
Trogons	Trogoniformes	1
Hornbills and Hoopoes	Bucerotiformes	4
Kingfishers and relatives	Coraciiformes	6
Woodpeckers and Toucans	Piciformes	9
Seriemas	Cariamiformes	1
Falcons and Caracaras	Falconiformes	1
Parrots	Psittaciformes	4
Passerines	Passeriformes	145

CHAPTER 2

Non-passerines

This group contains all birds that are not in the order Passeriformes. Unlike the passerines, they have no single distinguishing feature. For example, members of different orders may have four, three, or just two toes, and some birds may have feet that are webbed.

Order Struthioniformes

Families 1

Species 2

Size range 6–9 ft (1.8–2.8 m) tall

Distribution Sub-Saharan Africa

◄ On the lookout
With eyes 2 in (5 cm) wide and 6½ ft (2 m) above the ground, these female Common Ostriches (*Struthio camelus*) have a better view over the open plains than all other terrestrial animals except giraffes.

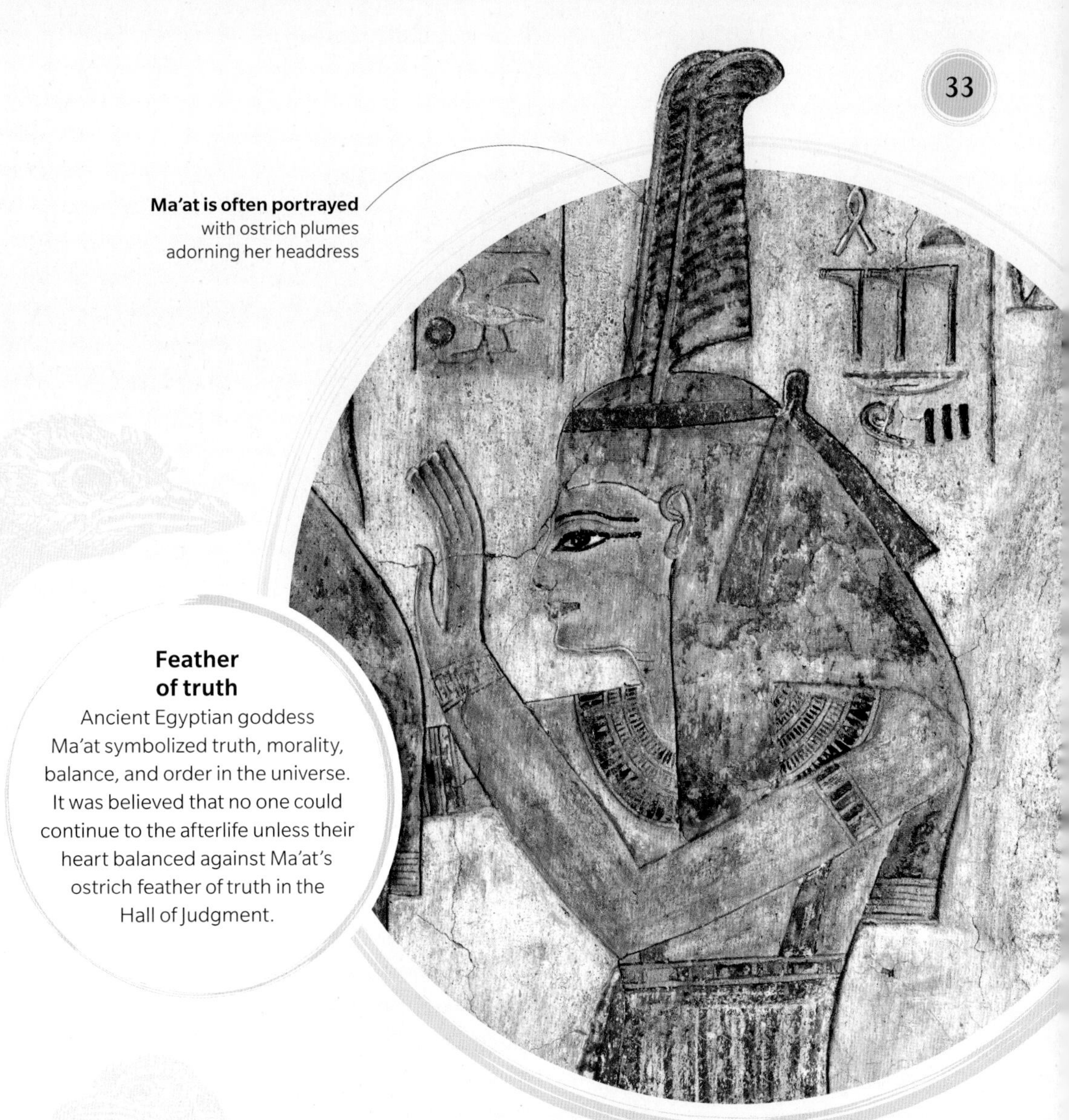

Ma'at is often portrayed with ostrich plumes adorning her headdress

Feather of truth
Ancient Egyptian goddess Ma'at symbolized truth, morality, balance, and order in the universe. It was believed that no one could continue to the afterlife unless their heart balanced against Ma'at's ostrich feather of truth in the Hall of Judgment.

Ostriches

Struthionidae

Astonishingly large and powerful, ostriches are visible at great distances across the open plains of Africa, while the distinctive, deep booming call of a dominant male carries far and wide.

Despite being the largest of all living birds, ostriches may fall prey to big cats, such as lions and cheetahs, and hyenas. They have also been hunted by humans for their meat and feathers for millennia; ancient Egyptian artworks depict the pharaoh Tutankhamun pursuing them with a bow and arrow. Unable to fly, their defense is to run, reaching speeds of 45 mph (70 kph), but they have a powerful kick, too.

Ostriches eat roots and seeds, and occasionally large insects, rodents, and reptiles, and swallow small stones to grind tough food. A common myth is that the ostrich will bury its head in the sand. This arose from distant sightings of ostriches reaching down to feed or to turn eggs during incubation. Early Christians thought the ostrich, through admirable concentration, could hatch its eggs by staring at them.

In the breeding season, males kick each other to establish the right to court and mate with females. Eggs are laid in a shallow nest scraped in the ground. Once hatched, groups of chicks from different broods may merge and be guarded by one or more adults for greater protection.

Eggs are decorated with charcoal or ash, sometimes to mark ownership

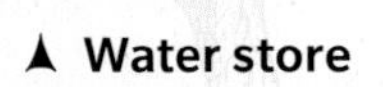

▲ Water store
For thousands of years, desert peoples in southern Africa have buried ostrich eggs filled with water to guarantee a drink during their hunting expeditions.

Order Apterygiformes

Families 1

Species 5

Size range 14–25½ in (35–65 cm) long

Distribution New Zealand

Kiwis

Apterygidae

Few birds are as emblematic of their country as kiwis. These unique flightless birds are recognized worldwide as a national emblem of New Zealand and its people are known as "kiwis."

▲ Probing bill
The North Island Brown Kiwi (*A. mantelli*) feeds by making characteristic holes in the soil with its long bill as it seeks out worms.

Found only in New Zealand, there are just five species of these chicken-sized birds, with the Okarito Kiwi (*Apteryx rowi*) being identified as recently as 2003. With hairlike plumage, no visible wings or tail, and tiny eyes, kiwis resemble mammals. These unusual birds have come to signify the uniqueness of the country and its wildlife. They have long been treasured by Māori and kiwi feathers were used to make cloaks for high-ranking people.

Burrowing in the forest

Kiwis are nocturnal, resting by day in hollows and burrows that they dig among the dense vegetation. Māori have long known these shy birds as the "hidden bird of Tāne," the god of the forest. Kiwis hunt on the forest floor, probing their slim bill into the litter and soil, and using their excellent sense of smell to find prey such as earthworms, beetles, and spiders. They also eat fruit and seeds. Long bristles at the bill base also help them detect prey by touch. Kiwis make loud whistles for contact and proclaiming territories, and in some species the call repeats up to ten times in a rising tone.

Kiwis form a lifelong monogamous pair bond, and the female lays an enormous egg—or occasionally two eggs—so large that there is no room for food in the stomach just before the egg is laid. In the Little Spotted Kiwi (*A. owenii*) and the North Island Brown Kiwi (*A. mantelli*), only the male parent incubates the eggs, doing so for 70–80 days. In the other three species, night-time incubation is shared by both parents. The young hatch fully feathered and are independent remarkably quickly, after only two weeks, and do not require feeding by the parent in that time.

COLOSSAL EGG

Kiwi eggs are distinctly disproportionate to the size of the mother. The huge egg can constitute 25 percent of a female's body weight and is six times the size of an egg of a similar-sized bird. The egg has a large yolk sac that nourishes the fully feathered young for two weeks after hatching.

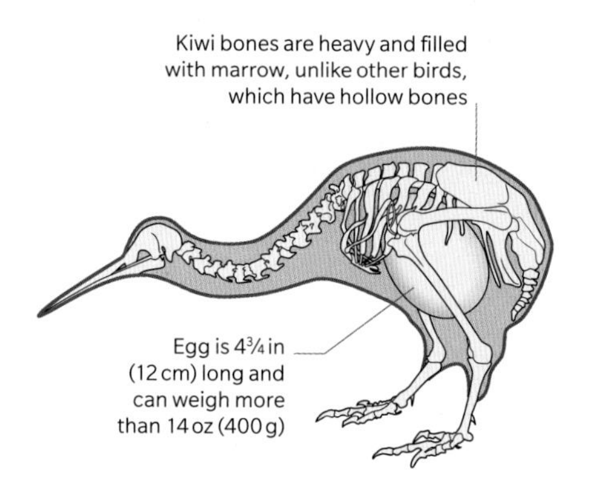

CROSS SECTION OF KIWI WITH EGG

Emu and Cassowaries

Casuariidae

Emus and cassowaries are the flightless giants of Australasia. With very loose, course plumage and colorful bare skin on the head and neck, these unusual birds are both captivating and fearsome.

Order Casuariiformes

Families 1

Species 4

Size range 3 ft 3 in–6 ft 3 in (100–190 cm) tall

Distribution Australia, New Guinea

Among the world's largest birds, the imposing Emu (*Dromaius novaehollandiae*) and cassowaries (*Casuarius* spp.) are tall and bulky, with a long neck and powerful legs for running and kicking, and three sharp-clawed toes on each foot.

Emus occur in open country in Australia where they feed on seeds, fruits, flowers, and other nutritious plants. Cassowaries are forest birds of Papua New Guinea and Australia that mainly eat fruits, including some that are poisonous to other animals, and they even swallow bananas whole.

Breeding booms

During the breeding season, emus and cassowaries make deep booming sounds that can be heard up to 1⅕ miles (2 km) away and even through dense forest. Emus lay 4–15 eggs and cassowaries lay 3–5 eggs. Males incubate the eggs and also rear the chicks. Emu egg laying has long been monitored by First Nations peoples of Australia using a star constellation called the Celestial Emu, which resembles a running emu. Its appearance in the sky in April coincides with the emu breeding season, and signals that the newly laid dark green eggs can be gathered for food. When the constellation appears horizontal in June, resembling a male sat on a nest, egg collection stops as embryos will have grown in the eggs.

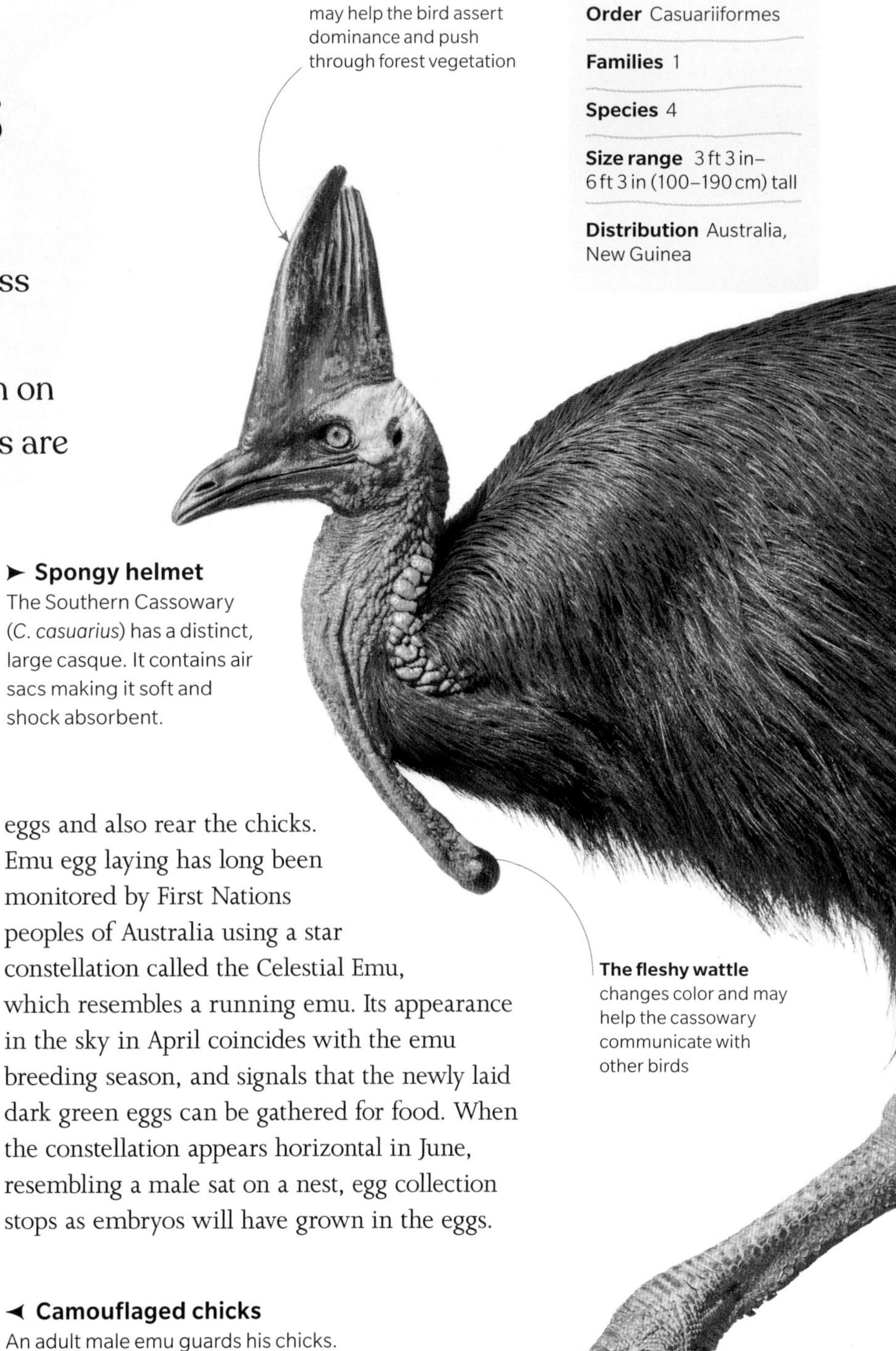

The distinctive casque may help the bird assert dominance and push through forest vegetation

➤ Spongy helmet
The Southern Cassowary (*C. casuarius*) has a distinct, large casque. It contains air sacs making it soft and shock absorbent.

The fleshy wattle changes color and may help the cassowary communicate with other birds

Sharply clawed toes aid digging for fallen fruit

◄ Camouflaged chicks
An adult male emu guards his chicks. The chicks are patterned with dark stripes, which camouflage them in open country and in shade.

Cassowaries can jump as high as 6 ft 7 in (2 m) straight up into the air thanks to their powerful legs

◄ Oriental finery
Several duck species have striking plumages, but none can compete with the male Mandarin Duck (*Aix galericulata*) for spectacular ornamentation and coloring. In contrast to its bold appearance, the bird itself is quite shy.

Rearing geese
This 14th-century version of *Tacuinum Sanitatis*, a handbook on health by Ibn Butlān, recommends living a life in harmony with nature.

Geese have been reared for eggs, meat, and feathers for thousands of years, and are also kept as "guard dogs"

Ducks, Geese, and Swans

Anatidae, Anseranatidae, Anhimidae

Order Anseriformes

Families 3

Species More than 170

Size range 12–71 in (30–180 cm) long

Distribution Worldwide except Antarctica

With a long history of domestication across the world, ducks, geese, and swans come second only to gamebirds in terms of their practical, economic, and cultural importance to humans.

Traditionally, the term "wildfowl" was used to refer to birds that were hunted for sport or food, and it became closely associated with ducks and geese. These are still the main groups of birds that are shot for sport without being reared and released. Domestic ducks, today raised for their meat, eggs, and down, derive from the Mallard (*Anas platyrhynchos*), domesticated in China more than 2,000 years ago, or from the mostly South American Muscovy Duck (*Cairina moschata*), reared by Indigenous populations for a similar period. Domestic geese, bred for the same purposes, can be traced back to the European Graylag Goose (*Anser anser*) or the eastern Asian Swan Goose (*A. cygnoides*). In fact, recent research suggests that geese may have been the first birds to be domesticated—in China some 7,000 years ago.

Ducks, geese, and swans vary greatly in shape, size, and coloration. The difference between a large white swan and a small brown duck is so stark that it is surprising to many people that they are grouped in a

Flat bill is hinged to facilitate pouring

► Ritual vessel
This Chinese bronze goose dating from the late Zhou dynasty is a container for wine, known as a tsun. Such vessels were used in ceremonial offerings to the dead.

single order, and almost all in the Anatidae family. Only four species are not: the three screamers, of South America, are in the family Anhimidae and the magpie goose of Australia and New Guinea is in the Anseranatidae.

Being closely associated with water, ducks, geese, and swans have waterproof feathers and short legs, and most species have webbed feet. Their feet spread wide on the backward stroke and close up on the forward one, to propel them efficiently through water. Most species have a rather broad but short bill. Others have a small hook to grip fish, and a few, known as sawbills, also have a serrated cutting edge.

Traits and tradition

Swans are among the heaviest of all flying birds: a full-grown Mute Swan (*Cygnus olor*) may weigh as much as 33 lb (15 kg). It can be difficult to tell the sexes apart, but in the Mute Swan, for example, males have a larger fleshy knob at the base of the bill. Swans form strong pair bonds, often for life, and are known to aggressively defend their territory. They have a long slender neck that helps them find food in shallow water, and they can also "upend" themselves, tilting the body vertically, to reach greater depths.

In the UK, swans have long been thought of as "royal" birds, and catching and killing them was a serious offense. An annual ceremony known as "swan-upping" has been performed since the 12th century, in which mute swans residing on the Thames River are marked on their bills (and then released unharmed). The tradition goes back to when swans were eaten at feasts, and marking them identified those that "belonged" to ancient trade associations or to the Crown.

Geese are smaller than swans, and have shorter necks. They spend the day feeding mostly on dryish grassland, arable land, or saltmarsh, and fly to safe areas of water or marsh to roost each evening. Where geese are numerous, large flocks become part of the local landscape. They make a loud chorus of contact calls, and fly in irregular masses, or arranged in long chevrons and "V" formations. Many, such as the Greater White-fronted Goose (*Anser albifrons*), are long-distance migrants, so they are associated with the changing seasons. Others, such as the Bar-headed Goose (*A. indicus*), can fly at very high altitudes.

▼ Hitching a ride
The Common Merganser or goosander (*Mergus merganser*) takes its chicks to the water right after they hatch. Unable to fly, the chicks have to leap from their nesting hole in a tree.

Diverse ducks

Ducks tend to be grouped into larger shelducks, surface-feeding or "dabbling" ducks, and diving ducks. There is much overlap, and many species

◄ **Formation flying**
Flying in a V-shaped formation allows these migrating graylag geese to save energy and communicate about orientation cues.

MALLARD
Anas platyrhynchos
This common dabbling duck is found in freshwater wetlands and coastal marshes. The female also has orange legs, but is brown with a bright blue patch on the wing.

MAGPIE GOOSE
Anseranas semipalmata
Found in wetlands in Australia and New Guinea, this primitive goose is unusually tall and has long toes, which are adapted for perching.

SOUTHERN SCREAMER
Chauna torquata
One of the three large South American screamers, this species has a crested head and hooked bill, and its wings have spurs for territorial combat.

" It doesn't matter if you're born in a duck yard, so long as you are hatched from a swan's egg! "

HANS CHRISTIAN ANDERSEN, *The Ugly Duckling*, 1843

use both fresh and salt water at times. Shelducks are gooselike in their stance and tendency to feed on mud or grassland, but they can be identified by their more strikingly patterned plumage. Diving species such as the Long-tailed Duck (*Clangula hyemalis*), scoters (*Melanitta* spp.), and eiders (*Somateria* spp.) breed on islands or beside lakes, but otherwise spend their time at sea, diving to feed on crustaceans and mollusks. Other diving ducks, such as the Tufted Duck (*Aythya fuligula*) and Canvasback (*A. valisineria*), similarly dive down from the surface, but mostly in fresh water. Dabbling ducks, such as the Northern

Bar-headed geese migrate over the **Himalaya** at heights of more than **20,000 ft** (6,000 m)

Shoveler (*Spatula clypeata*) and Eurasian Teal (*Anas crecca*), feed by sieving food from water or wet mud. Their bills have a fine network of comblike projections (lamellae) to filter solids from water, which they then manipulate using the tongue. Other surface-feeding species, such as the Eurasian Wigeon (*Mareca penelope*), graze on marshland and meadows.

Male (drake) ducks are more colorful and boldly patterned than females and juveniles, and most have their finest feathers during the winter. Wearing their "courtship plumage," males perform ritualized movements beside one or more females in the hope of forming a pair. They also perform courtship chases in flight.

After mating, and once they have moved to the breeding area (often involving long-distance migration), the females incubate the eggs and care for the young, while the males molt into a duller "eclipse" plumage. This involves shedding most of their flight feathers at the same time, meaning that eclipse ducks (and some geese) become flightless for a short period.

Duck nests are lined with down from the female's plumage. They may be built on the ground among grasses, in a cavity, or (as with shelducks) in a rabbit hole or a hole in a tree. Chicks are covered in down, and are active very soon after they hatch. They quickly begin to swim and forage for themselves.

> "The swan, like the soul of the poet, By the dull world is ill understood."
>
> HEINE HEINRICH, *The Poems of Heine: Complete*, 1866

➤ **Hidden feathers**
The slender-necked Australian Black Swan (*C. atratus*) appears to have all-black plumage until it takes flight and reveals bold white wing tips.

The first European to see a **black swan** was probably Antonie Caen in **1636** in Western Australia

▲ Defending her territory
Dutch artist Jan Asselijn painted *The Threatened Swan* in c.1650. It was later interpreted as a political allegory, and may represent Dutch politician Johan de Witt protecting his country throughout a number of conflicts.

Some Indigenous populations of North America, such as the Algonquian, recount legends about gullible ducks and geese being tricked so they could be caught and consumed. However, in Greek mythology, the duck is a symbol of fidelity and caring for the family. The Greek goddess Penelope, renowned for her loyalty to her husband Odysseus, is believed to have been named after the Greek word for "duck," because she was rescued from the sea by a family of ducks after her father discarded her for not being a boy.

ARCTIC ROBES

Two Tlingit women of southern Alaska pose for a photograph, wearing traditional robes made of marmot furs and eider duck feathers or down. Like other ducks, eiders line their nests with their own down for insulation. They pluck the lightweight feathery down from beneath their tough exterior feathers. Eider down has been harvested as a warm filling for quilts and pillows for centuries. It is only taken after the ducklings vacate the nest so the birds are not harmed in the process.

STUDIO PORTRAIT, C.1900

Heading north

Having spent the winter in northwestern Europe, Pink-footed Geese (*Anser brachyrhynchus*) migrate to their Arctic breeding grounds in spring. Geese tend to follow the same routes every year, and thousands of them break their journey to roost on pastureland in central Norway, but this flock was met with still wintry conditions in the fields where they feed.

Badge of honor
Embroidered rank badges were awarded throughout China's Qing Dynasty. The golden pheasant denotes a second rank civil official.

Pheasants

Phasianidae

This large family of strong-legged, ground-dwelling gamebirds is of immense and unrivaled economic and cultural importance across most of the world.

In terms of economic significance, no other bird family comes close to the pheasants. One member, the Red Junglefowl (*Gallus gallus*), is the primary ancestor of the domestic chicken. Today, there are at least 33 billion chickens being reared for their meat and eggs, making them by far the world's most numerous bird. The pheasant family includes some of the most well-known gamebirds —birds that are hunted and killed for sport as well as food (the term is also used for all birds of the order Galliformes). The Common Pheasant (*Phasianus colchicus*) has been hunted and eaten in quantity in many countries for centuries, as have quails (*Coturnix* spp.). Some family members provide a source of local bushmeat, including, in Africa, the many species of francolin. In Tanzania, the Udzungwa Forest Partridge (*Xenoperdix udzungwensis*) was only recently discovered by Western ornithologists when its distinctive feet were found at the bottom of a cooking pot at an expedition camp in the Udzungwa mountains.

The Maya are thought to have domesticated the Wild Turkey (*Meleagris gallopavo*) around 2,000 years ago. Today, eating roast turkey at Christmas and Thanksgiving celebrations is an established Western tradition. In other parts of the world, however, instead of being eaten some pheasant family members are revered or worshipped. The

Order Galliformes

Families 1

Species More than 180

Size range 4¾–94 in (12–240 cm) long

Distribution
Worldwide except South America and Antarctica

Indian Peafowl (*Pavo cristatus*), popularly known as the "peacock," is deeply embedded in Hindu tradition. Hindu deities are commonly shown riding on peacocks, and Lord Krishna is most often depicted with a peacock feather in his hair, symbolizing humility, purity, and devotion, among other qualities. Across much of India, peacocks are a common sight in villages and farms. Their loud cries are often heard at tourist sites, and their feathers are sold to visitors as souvenirs. The White Eared Pheasant (*Crossoptilon crossoptilon*), found around the Himalaya, is sacred to Buddhists in the Sichuan province of China, where protected habitats are provided.

Native to much of the world, members of the pheasant family vary considerably in size, but they are all plump, small-headed birds, with strong legs and feet, and rounded wings. They mostly live on the ground, although many species, including pheasants, turkeys, and peafowl, roost in trees at night if they can. The Rock Ptarmigan (*Lagopus muta*) nestles into the snow at night and may dig a burrow. Grouse are unusual in that they often spend a long time feeding in trees, sometimes for months on end. Most members of the pheasant family have three long, forward-facing toes, with sharp claws and a small hind toe. Males often have at least one

▲ Golden choice
Competition for females is high among China's brilliantly colored male Golden Pheasants (*Chrysolophus pictus*) in the breeding season, and many remain unmated.

"spur" facing backward, above the hind toe, which is used in battle and varies in size—a clue for females as to the quality of the male. The feet are well adapted for walking and running, which are the preferred methods of escape for most family members, and also for scratching the ground to find food, including grains, seeds, fruits, leaves, insects, and small mammals and reptiles. The Indian peafowl is known to eat snakes smaller than some 8 in (20 cm) long.

On the move

All members of the pheasant family can fly but many do so reluctantly, often as a last resort to escape predators. They have strong flight muscles that power their heavy bodies as they accelerate rapidly and bolt away at high speed. These well-developed flight muscles and the overall chunkiness of their main body are part of the reason why these birds make such good eating. During flight, they perform bursts of wingbeats followed by glides. However, they soon tire and land quickly.

A few species migrate long distances, including the Common Quail (*C. coturnix*) and the Harlequin Quail (*C. delegorguei*). The former, identified by some as the food that God sent into the desert in the Bible, may move twice a year. It is thought that birds from North Africa migrate to Europe, breed, migrate again (with the offspring), and breed again farther north. In contrast, Sooty Grouse (*Dendragapus fuliginosus*) may "migrate" only 1,300 ft (400 m) downhill for the winter.

The pheasant family shows a wide range of breeding behavior, from entirely monogamous, as in the Gray Partridge (*Perdix perdix*), to wildly

◄ Not satisfied
Gustave Moreau's 1881 watercolor depicts a beautiful peacock complaining to the Roman goddess Juno that he does not have the lovely voice of a nightingale. She scolds him, saying he is clearly already favored enough.

Strutting his stuff
Male Sage Grouse (*Centrocercus urophasianus*) splay their feathers and inflate bulbous sacs on their chest during courtship.

polygynous, as in various species of grouse and pheasant. Males gather in close proximity in leks (or within hearing range in "exploded leks") so that females need only attend briefly to select and then mate with their preferred male. Communal courtship rituals are especially prevalent among grouse, and males compete vigorously for attention and dominance, often fighting and posturing wildly. In North America, a number of Indigenous dances and ceremonies are based on their antics.

Perhaps the most familiar courtship display is that enacted by the peacock, which spreads its ornate upper tail covert feathers into a large eye-catching fan. The Great Argus (*Argusianus argus*) also makes an all-round fan, which is mainly brown with white "eyes." Many species advertise with loud calls and may make extraordinary sounds, such as the "champagne cork pop" of the Western Capercaillie (*Tetrao urogallus*). Several pheasants have remarkably long tail feathers, notably the Reeves's Pheasant (*Syrmaticus reevesii*), which has the longest in the world, at more than 8 ft (2.4 m) in length.

All birds in the pheasant family build their nest on the ground, usually no more than a scrape in the soil in a discrete location in hedgerow or long grass. Several species lay notably large clutches, including the gray partridge, which lays 15–17 eggs. Red-legged Partridges (*Alectoris rufa*) may lay two clutches, with each member of the breeding pair incubating one.

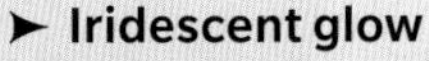

Metallic green wiry feathers comprise the male's striking crest

➤ Iridescent glow
The male Himalayan Monal (*Lophophorus impejanus*) has striking plumage. Both sexes are found at altitudes of up to 14,800 ft (4,500 m).

> "A wind from the Lord sprang up and brought quail from the sea . . ."
>
> *The Bible,* Numbers 11: 31

ROCK PTARMIGAN
Lagopus muta
One of the world's hardiest birds, this species occurs as far north as 83°N in Canada. Its feathered feet provide extra insulation when walking on snow.

RING-NECKED PHEASANT
Phasianus colchicus
Native to parts of Asia, it was introduced to Europe and North America and has been widely hunted there for centuries.

WILD TURKEY
Meleagris gallopavo
The turkey is the only widely domesticated North American bird. Formerly greatly depleted, populations have been restored in recent years over much of its native range.

Guineafowl and New World Quails

Numididae, Odontophoridae

Order Galliformes

Families 2

Species More than 40

Size range 6¾–28 in (17–72 cm) long

Distribution North and South America, Africa

Guineafowl are distinctive plump birds that sometimes live in large groups, while the smaller New World quails are more secretive and shy.

The guineafowl of Africa have bare skin on the face and neck, and head adornments such as crests and helmets. They feed on the ground on seeds and plants, with some insects and small mammals taken, too. The Helmeted Guineafowl (*Numida meleagris*) has been domesticated around the world. It was known to the ancient Greeks and Egyptians, and the myth of Meleager links the bird's white-spotted plumage to the teardrops of his grieving sisters. A noisy bird, it is often seen alongside big game in Africa, and associates with monkey troops, watching for danger.

Jagged edge

New World quails differ from the many similar small gamebirds by the serrated sides of the lower mandible of the bill. These help them deal with plant matter, including tearing leaves. Many scratch the ground litter with their strong legs, and also dig. Some live in forests, while others are notably tolerant of arid habitats. While New World quails are ground-living and well known for their explosive takeoff, they often perch up high in trees or cacti.

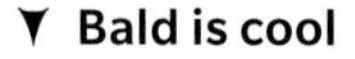

▼ Bald is cool
Named for its bare head, the Vulturine Guineafowl (*Acryllium vulturinum*) lives in arid regions of East Africa.

Bare skin on the head and neck aids cooling

Blue-tinged plumage with white spots or plumes is typical of most guineafowl

◄ Handsome topknot
This engraving of a California Quail (*Callipepla californica*) from 1849 shows off its forward-curving head plume.

Chachalacas, Curassows, and Guans

Cracidae

Order Galliformes

Families 1

Species More than 50

Size range 16½–37 in (42–95 cm) long

Distribution Southern Texas, Central and South America

Found from southern Texas to northern Argentina, these distinctive gamebirds are known for their unusually arboreal lifestyle, nesting in trees rather than on the ground, and their noisy vocalizations.

Chachalacas, curassows, and guans spend more time in trees than other gamebirds. They have a long, well-developed hind toe that gives them extra stability when perching, and also helps them run along branches in agile fashion. Other typical features include a long tail and a very small head, albeit often with an adornment such as a crest or wattle. Found in tropical and subtropical Central and South America, many species live in thick forests, where fruit is always an important part of their diet.

The birds in this family are known for their loud calls, and the name "chachalaca" describes well the squawking offerings of this group. Some guans have an adapted trachea that curves around the pectoral muscles, amplifying their crowing or braying calls. They also have adapted wing feathers that produce rattling sounds when flapped. All species in this family have rounded wings, but the large curassows spend more time on the ground and are poor fliers.

Head adornments such as this impressive crest are seen in many species of curassows

Long, strong legs are used for running and scratching the ground

The long tail helps the bird balance

➤ **Blending in**
This female Great Curassow (*Crax rubra*) shows typical cryptic plumage, which helps hide her from predators when she remains still.

PLAIN CHACHALACA
Ortalis vetula
This very noisy species lives in the scrubby, thorny forests of Central America, and is the only chachalaca that is also found in southern Texas.

DUSKY-LEGGED GUAN
Penelope obscura
This sociable forest species feeds both up in the trees and on the ground, especially on fruit. It builds a bowl-shaped nest out of twigs, protected by dense cover.

▲ **Egyptian water jar**
This clay water jar (c.3850–2960 BCE), from a civilization that predates the First Dynasty in Egypt, is decorated with a water pattern and flamingos, indicating that the birds held some cultural significance.

> "... a dish of flamingo's tongues being fit for a prince's table."
>
> WILLIAM DAMPIER, *A New Voyage Around the World,* 1697

Order Phoenicopteriformes

Families 1

Species 6

Size range 31½–57 in (80–145 cm long)

Distribution North America, South America, Southern Europe, Middle East, Africa, South Asia

Flamingos

Phoenicopteridae

With a long sinuous neck, stiltlike legs, and striking pink plumage, flamingos have captured the imagination for millennia and are even associated with the Phoenix of Western mythology.

The pigments that give all flamingos their reddish plumage are derived from carotenoids in their food, and the intensity of color is related to the quality of their diet. Flamingos are filter feeders and their bill is adapted for the task of sieving anything from microscopic diatoms and microalgae to crustaceans, mollusks, aquatic insects, annelid worms—even small tadpoles and fish—to plant matter such as seeds.

Flamingos breed in colonies and their nest is a mud cone in which a single egg is laid. Colonies rarely comprise fewer than 50 pairs and the largest can contain over a million nests, as in the case of Lesser Flamingo (*Phoeniconaias minor*). Its largest population—on salt lakes in the Rift Valley of East Africa—comprises 1.5–2.5 million individuals.

The Phoenix myth

The common name for this family is derived from Portuguese or Spanish, and has the same origin as the word "flame," while the root of the genera *Phoenicopterus*, *Phoeniconaias*, and *Phoenicoparrus*, meaning "flame-colored," is the same as that of the mythical Phoenix. Beyond the reddish color, the reasons for this association are unclear, but since perhaps as far back as the time of ancient Greek historian Herodotus, some 2,500 years ago, the characters of flamingo and Phoenix have been intertwined. One theory suggests that in the shimmering African heat, a bird on its mud nest may appear to be rising from still-warm ashes.

FILTER FEEDING

Flamingos are filter-feeders and are the only birds that feed with their bill held upside down. They use their strong tongue to rapidly pump water in and out of their bent bill and sieve out food in the process.

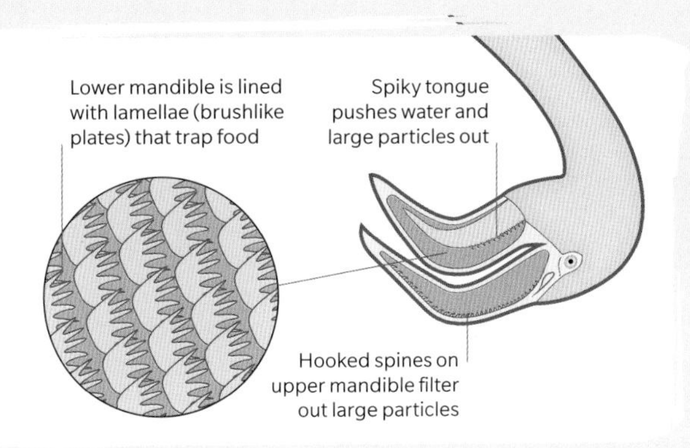

CROSS SECTION OF BILL

▼ Brilliant pink
The American Flamingo (*Phoenicopterus ruber*) is the most intensely colored of the flamingos. Chicks and juveniles are gray but will acquire pink plumage as adults.

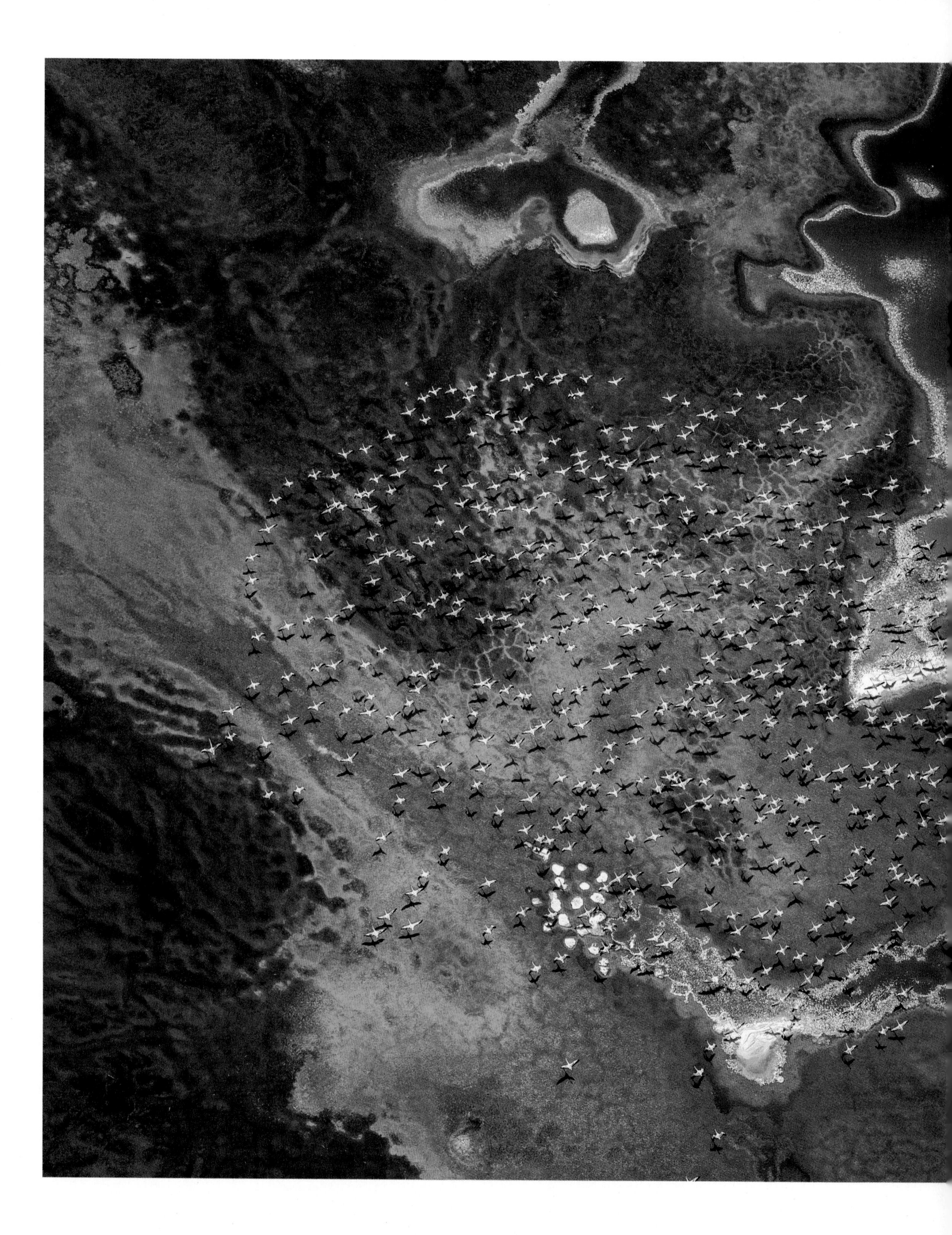

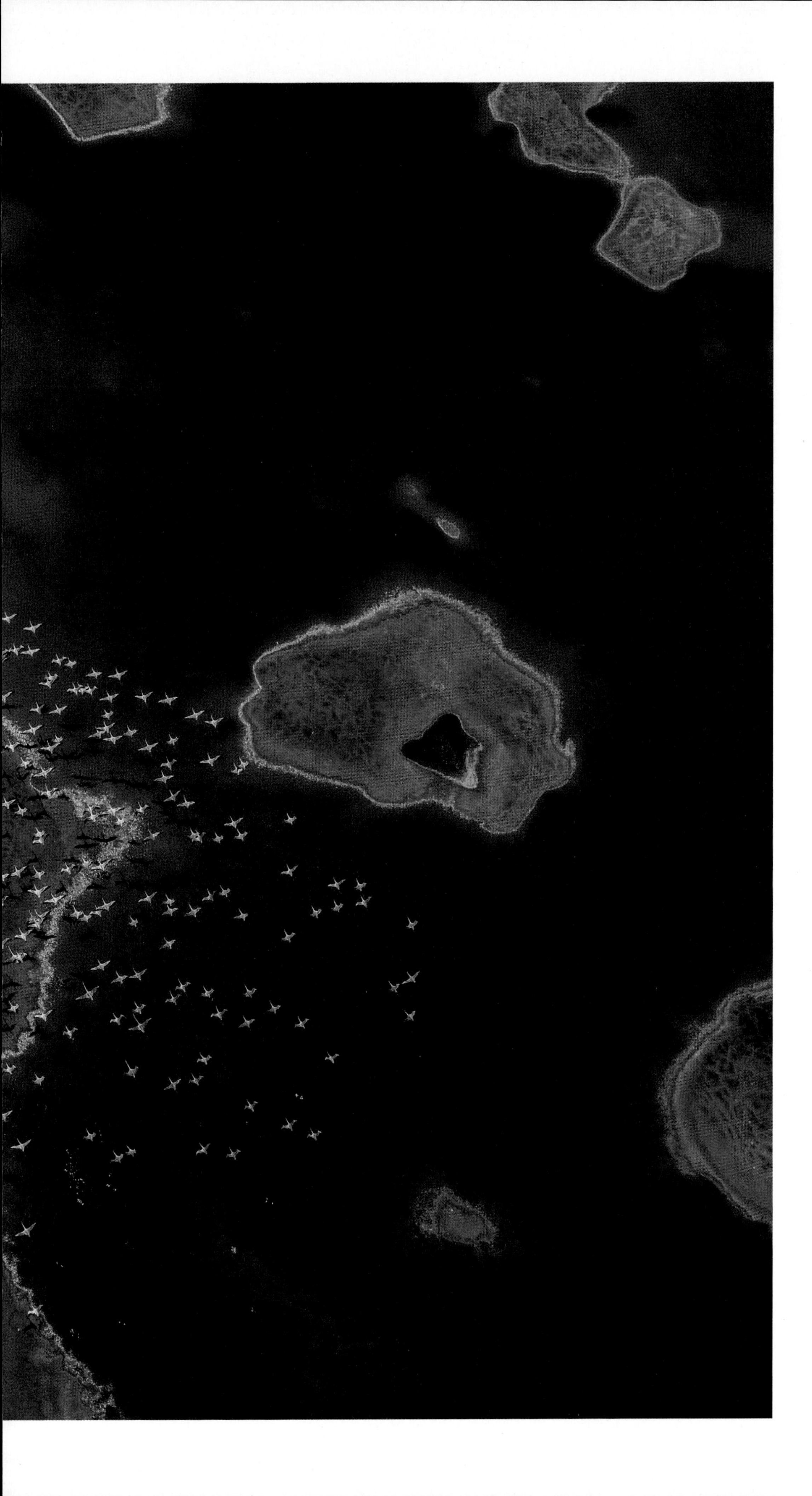

Soda lake specialists
Lesser flamingos flock in huge numbers to the soda lakes in East Africa's Great Rift Valley to feed on the cyanobacteria that thrive in the alkaline waters. The pigments in this food turn their plumage pink. Flying with their long legs and neck outstretched, they cast distinctive shadows on the water below in the bright sunlight.

Long central streamers emerge from a shorter tail, giving the tropicbird a graceful shape in flight

Order Phaethontiformes

Families 1

Species 3

Size range 30–40 in (76–102 cm) long

Distribution
North and South America, Africa, Asia, Australasia

Black wing markings are typical of all three tropicbird species

Tropicbirds

Phaethontidae

▲ Bosun bird
The White-tailed Tropicbird (*Phaethon lepturus*) and its relatives were named "bosun birds" by sailors due to the tail resembling a marlinspike and their calls, which are similar to a ship's bosun's whistle.

True to their name, tropicbirds spend most of their life soaring above the tropical waters of the Atlantic, Pacific, and Indian Oceans.

With a stocky but elegant shape, a short tail adorned with very long streamers, and a thick bill, a tropicbird shines white against the blue sea. Tropicbirds breed in colonies on rocky islands, nesting in the shelter of tree roots or rock cavities. Groups of birds perform spectacular courtship flights, gliding above one another, but they are solitary for the rest of the year.

Tropicbirds often catch flying fish skimming just above the waves, but they plunge-dive for other fish and squid. Air sacs under the skin protect the bird's head on impact and give tropicbirds their thick-necked shape.

Pacific island fishermen have long prized tropicbird feathers as lures, believing that only feathers plucked from a live bird stay lustrus. In the early 20th century, a fashion for elaborate feather hats saw tropicbirds hunted and traded, which reduced the size of many colonies.

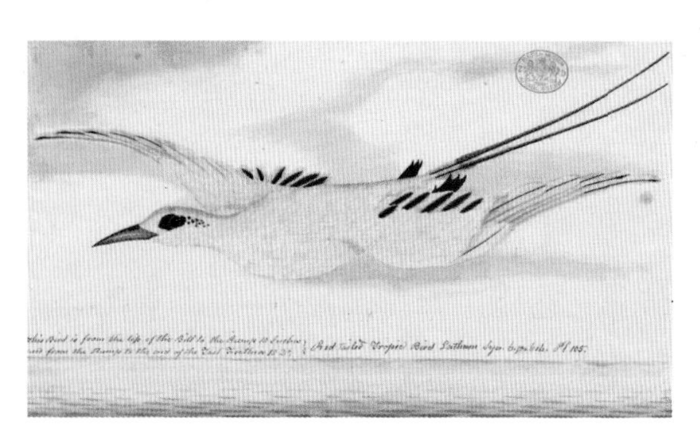

◄ Eye-catching streamers
The vivid Red-tailed Tropicbird (*Phaethon rubricauda*) was portrayed in the late 18th century by an artist known as the Port Jackson Painter, who may have sailed to Australia with the First Fleet.

RED-BILLED TROPICBIRD
Phaethon aethereus
Similar in appearance to other tropicbirds and common in the Caribbean, this species is identified by its red, faintly curved bill and white tail streamers.

Kagu

Rhynochetidae

Order	Eurypygiformes
Families	1
Species	1
Size range	22 in (55 cm) long
Distribution	New Caledonia

Widespread across New Caledonia until the arrival of Europeans in the late 18th century, the elusive kagu is now found only in forests in the south of the island.

Loud barking calls at dawn draw attention to the Kagu (*Rhynochetos jubatus*), but this light gray bird is otherwise inconspicuous—locals call it the "ghost of the forest"—except when the crest is raised and wings spread to reveal dark bars in displays, or as it scrambles over rough terrain. Fleshy flaps over the nostrils help keep them free of soil when the bird probes for food such as worms, reptiles, and insects.

Struggle for survival

Habitat loss and predation by cats, dogs, and rats introduced to New Caledonia by Europeans has led to this flightless bird becoming endangered. There are now fewer than 2,000 kagus, largely confined to two forest parks with scattered pairs elsewhere. Most are immature and almost 60 percent are male, making the maintenance of a viable population unusually difficult. However, local conservation efforts have seen kagus successfully reintroduced into a park where predators are controlled.

Silver crown
The chief feature of a mature kagu is its high, fanlike crest, which is held erect in courtship displays. Its deep red eyes and bright orange bill stand out against its otherwise pale appearance.

The name kagu comes from the Kanak people and is thought to mimic the bird's unusual call

Wings are flapped like a cape during displays

▲ Special issue
New Caledonia postage stamps celebrate the kagu's courtship display, in which it spreads its wings to reveal striking dark bars.

▼ Crowning glory

In a family notably lacking ornamentation, four large species of crowned pigeons, including the Victoria Crowned Pigeon (*Goura victoria*), are exceptions. These elaborately crested birds forage on the forest floor in New Guinea.

Order Columbiformes

Families 1

Species More than 340

Size range 9–27 in (22–70 cm) long

Distribution Worldwide

The dove has been a symbol of peace around the world for thousands of years

Divine messenger
The Biblical description of the Holy Spirit descending like a dove is often represented literally in Christian art. In 1460, Giovanni Francesco da Rimini depicted God, with four angels, releasing a white dove that represents the Holy Spirit, carrying God's word.

Pigeons and Doves

Columbidae

Familiar around the world, these soft-plumaged and small-headed birds have a slim bill, with a bare, fleshy patch at the base, and short legs that give them a waddling walk.

Feeding on fruit, seeds, grain, and other vegetable matter, pigeons are mostly found in forests and scattered woodland, on farmland, and in suburban areas. Wild Rock Doves (*Columba livia*) were domesticated for food, especially the fat nestlings, or squabs, at least 5,000 years ago, (and the first dovecotes were built in the Middle East around 4,000 years ago). Their descendants became the source of all racing and homing pigeons, and some escaped to become the feral pigeons seen in towns and cities worldwide.

The remarkable homing ability of these birds to navigate across long distances to their home has been used for at least 3,000 years. The Romans used carrier pigeons to report the results of chariot races, and in Asia "pigeon post" services were established, with a pigeon being transported to the source of the message and then flying back home. This ability has also been exploited in pigeon racing, and was important in the 20th century when pigeons carried vital messages and news during both World Wars.

Domestic pigeons can **navigate their way home** from as far away as **1,300 miles (2,000 km)**

▲ **Powerful launch**
The stocky, deep-breasted shape of the wild rock dove reveals strong muscles that give a powerful, rapid rise into the air and stamina for long-distance flight.

Most birds can navigate to a degree, and long-distance migrants make use of the Earth's gravitational field and the positions of the sun and moon to find their way across oceans and continents. Racing pigeons are exercised in flights around their home lofts and learn to recognize local landmarks. They navigate from a distance using magnetic fields detected in their beak, then identify the loft by sight and smell.

Clapping wings

Pigeons and doves are unusual in that most have no flight or contact calls. Instead, many communicate using the sound of their wings clapping together as they fly off. Some, such as the Common Wood Pigeon (*Columba palumbus*), have distinctive high-rising display flights. Their songs are repetitive but recognizable patterns of cooing and purring notes. Some of these calls are highly characteristic and evocative of certain areas, such as the short, descending sequences of various wood doves (*Turtur* spp.) that are commonly heard in the African savannas.

HUNTED TO EXTINCTION

In the mid 19th century in North America, migrating hordes of Passenger Pigeons (*Ectopistes migratorius*) numbered hundreds of millions, taking hours to pass by, thus making them ideal targets for hunters. Their reliance on gigantic flocks to outwit natural predators made them vulnerable to the humans that burned and poisoned as well as shot them. By 1890, flocks were reduced to dozens of birds and the last captive bird died in 1914.

LOUISIANA, US, c.1870s

LUZON BLEEDING-HEART PIGEON
Gallicolumba luzonica
Restricted to a single Philippine island, this is one of five bleeding-heart species known for a vivid red mark on the breast.

COMMON WOOD PIGEON
Columba palumbus
Abundant in farmland and woods in most of Europe, most populations are resident but those in the north and east move south and west in winter.

NAMAQUA DOVE
Oena capensis
One of the smallest doves, this species' long tail is eye-catching as it flies up from the ground in semiarid regions of Africa and the Middle East.

Sandgrouse

Pteroclidae

Order	Pterocliformes
Families	1
Species	16
Size range	9–16 in (22–41 cm) long
Distribution	Southern Europe, Africa, Asia

Found in arid places, sandgrouse are hardy ground-dwelling, swift-flying birds with cryptic camouflage, making them very hard to see except when they visit waterholes to drink.

With the plump body of a gamebird but the narrow wings of a wader, sandgrouse are birds of dry scrubland, grasslands, and deserts, where they feed almost exclusively on seeds. Living in open habitats with little cover in which to hide from predators, they have highly cryptic plumage, often with spots and lines to break up their shape. Their short legs are feathered, and a complete coating of inner down provides insulation in extreme temperatures.

Sandgrouse are best known for their habit of visiting pools to drink communally, sometimes in the thousands. This daily mass gathering usually takes place at dawn, but in some species, such as Lichtenstein's Sandgrouse (*Pterocles lichtensteinii*), it takes place at night. Sandgrouse are fast fliers and travel great distances from their nests—scraped into the ground—in search of water.

➤ Hydrating feathers
Male sandgrouse have specially adapted spongelike breast feathers. After soaking, the parent flies up to 30 miles (50 km) to provide its young with water, drunk directly from the feathers.

▼ Life partners
Pallas's Sandgrouse (*Syrrhaptes paradoxus*) mate for life, breeding in the vast steppes of central Asia. The male is similar to the female except for a bold orange face.

Order Otidiformes

Families 1

Species More than 20

Size range 16–59 in (40–150 cm) long

Distribution Europe, Asia, Africa, Australia

Bustards

Otididae

Long-legged and ground-living, bustards inhabit open semidesert, grassland, or cropland areas, and are very sensitive to disturbance.

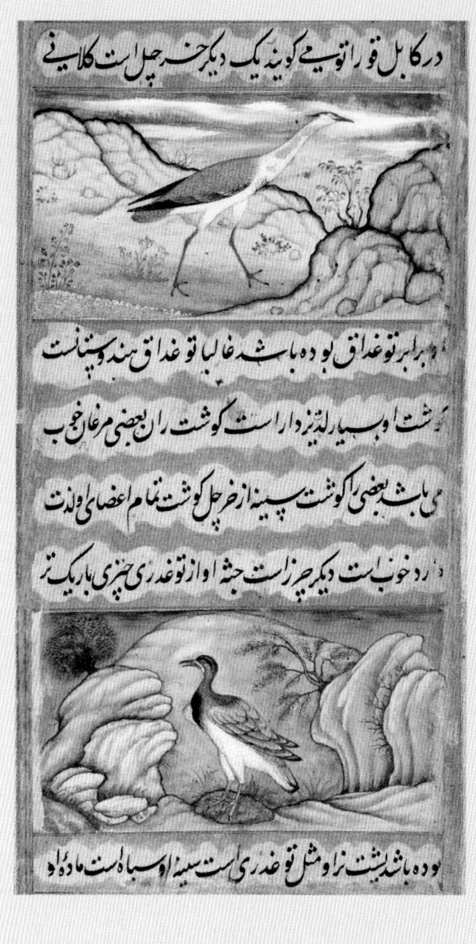

▲ Memoirs of Babur
Dating from the late 1500s, this illustrated copy of the Mughal emperor's journal depicts a Great Indian Bustard (*Ardeotis nigriceps*) top, and a now critically endangered Bengal Florican (*Houbaropsis bengalensis*).

With several species being so large, striking, and good to eat, bustards have always had a difficult relationship with humans. Hunting and trapping still take place in Iran and Pakistan, for example, but in most areas it is habitat loss that pressurizes populations. Smaller bustard species known as korhaans still thrive in African savannas, but these once-pristine expanses are shrinking. In southern England, the Great Bustard (*Otis tarda*) disappeared some 200 years ago, but reintroduction schemes begun early this century are succeeding.

Male bustards gather to perform conspicuous courtship displays in "leks" (communal areas), and they attract females over great distances. They strut, twirl, and contort their bodies until selected. Occasionally, fights break out. After mating, the male continues to dance to attract his next female. No pair bonds are formed, and females incubate their eggs alone in a shallow scrape in the ground.

◄ Puffball pose
Turning much of its plumage almost inside out, a male great bustard marches in a dramatic display of courtship, revealing extensive white plumes and an inflated blue-gray throat.

KORI BUSTARD
Ardeotis kori
One of the world's largest and heaviest flying birds—weighing up to 42 lb (19 kg)—this African species has a wingspan of up to 2½ ft (76 cm).

AUSTRALIAN BUSTARD
Ardeotis australis
This bustard is thriving in northern and inland Australia, but nonnative predators, such as foxes, threaten it elsewhere. When disturbed, it tends to stalk away slowly.

LITTLE BUSTARD
Tetrax tetrax
The Little Bustard is a sociable bird. The male inflates its neck, flicks back its head, and creates a short, dry rattle with its bill during courtship.

➤ Toe hold
Native to Angola, the Red-crested Turaco (*Tauraco erythrolophus*) lives in the foliage of forest trees. Thanks to their unusual toe arrangement, turacos are exceptionally agile when maneuvering between trunks and branches.

GREAT BLUE TURACO
Corythaeola cristata
The largest and most widely distributed of the turacos, this species is found in forests from West to Central Africa. It is sociable, and several pairs may nest in one tree.

GREY GO-AWAY-BIRD
Crinifer concolor
This is one of three species that live in more open habitats. Its name derives from its bleating call, which some hunters say is made to warn game animals of their presence.

The distribution of turacos corresponds with one of the world's richest copper belts

➤ **Wood-carved mask**
This example of a male *chihongo* face mask from the Chokwe people of Angola dates from 1850–1900. It was used in initiation rites and was believed to confer esteem, courage, and wisdom.

Order Musophagiformes

Families 1

Species More than 20

Size range 16–30 in (40–75 cm) long

Distribution Africa

Turacos

Musophagidae

Familiar in sub-Saharan Africa but found nowhere else in the world, turacos have unique plumage and no close relatives. Most species live in extended family groups and make remarkably noisy calls.

Turacos are medium-sized birds of forests and bushland. Their eye-catching colors are particularly fascinating to scientists because their plumage has two pigments that, uniquely, are copper-based. Turacin is a bold red pigment found in the wings of several species, while turacoverdin is the green that suffuses much of the plumage of the forest turacos. Both pigments are acquired from the birds' diet of fruit, and ornithologists estimate that an adult needs to consume 45 lb (20 kg) of fruit to ingest enough copper to keep the colors. Remarkably, for such a rare specialization, not all the family members exhibit both pigments, and the go-away-birds (*Crinifer* spp.) show none at all.

In some parts of Africa, brilliantly colored turaco feathers are used as a sign of status. For example, in Cameroon, a red flight feather from Bannerman's Turaco (*Tauraco bannermani*) in a man's hand-woven hat indicates his position as a council member. And the Zulu king Cetshwayo, who ruled from 1872 to 1879, was said to be greatly protective of feathers from the Purple-crested Turaco (*Gallirex porphyreolophus*), not permitting anyone else to wear them.

◄ **Conspicuous crest**
This hand-colored steel engraving from 1837 depicts the Guinea Turaco (*Tauraco persa*), whose bright green crest opens up to form a semicircle.

◄ Irresistible urge
A Common Reed Warbler (*Acrocephalus scirpaceus*) continues to feed a common cuckoo chick even when it is more than three times its own size. The cuckoo fledgling's repeated begging call tricks the host parent into feeding it more than its own offspring.

Order Cuculiformes

Families 1

Species 150

Size range 6–31½ in (15–80 cm) long

Distribution North and South America, Africa, Europe, Asia, Australasia

Cuckoos

Cuculidae

Cuckoos are elusive so most people may never see one, but they are, nevertheless, well known for laying their eggs in other birds' nests.

Cuckoos are widely associated with leaving their chicks to be raised by host parents. Yet two-thirds of all cuckoo species build their own nests and rear their own young. Those that are brood parasites of other birds may lay eggs of a similar color to those of their host, but the Common Cuckoo (*Cuculus canorus*) is known to sometimes lay a brown egg among a clutch of blue ones.

Cuckoo calls

Most cuckoos are heard more often than they are seen. The eponymous "cuc-koo" call is famous, but only one species—the common cuckoo, herald of the European spring—actually sings "cuc-koo." Other species' calls include whistling, fluting, or bubbling sounds. In Japan, where cuckoos are associated with the coming summer, their song is said to represent the dead desperate to rejoin their loved ones. Greek mythology considers the cuckoo to be the sacred bird of Hera, goddess of marriage, and king of the gods Zeus transformed into a cuckoo in order to woo her. In Indian myths, the Jacobin Cuckoo (*Clamator jacobinus*) heralds the arrival of the monsoon season, while the Asian Koel (*Eudynamys scolopaceus*) and Indian Cuckoo (*Cuculus micropterus*) often symbolize the welcome dawn of a new year, fertility, and hope.

A long bill with a hooked tip enables the roadrunner to keep a firm grasp on its prey

▲ Fast food
Up to 24 in (60 cm) long, this large American cuckoo, the Greater Roadrunner (*Geococcyx californianus*), can run at speeds of around 20 mph (30 kph), perhaps faster in short bursts, when threatened or in pursuit of reptilian prey.

◄ Cuckoo and azaleas
Renowned Japanese artist Katsushika Hokusai depicts a Lesser Cuckoo (*Cuculus poliocephalus*) in this ukiyo-e print from 1834. Despite the open bill in this flying bird, this cuckoo gives its "hoo-hoo" song from a perch.

ASIAN EMERALD CUCKOO
Chrysococcyx maculatus
Glistening green (but browner on the head of females) and broadly barred beneath, this is one of several species of small (7 in/18 cm long) tropical, forest-dwelling cuckoos with vivid coloring.

PHEASANT COUCAL
Centropus phasianinus
The long tail and close barring of brown, cream, rufous, and black give rise to this cuckoo's common name. This mostly ground-feeding species is found in Australia, New Guinea, and East Timor.

Nightjars, Frogmouths, and Potoos

Caprimulgidae, Podargidae, Nyctibiidae

Orders Caprimulgiformes, Podargiformes, Nyctibiiformes

Families 3

Species 120

Size range 7–23 in (19–58 cm) long

Distribution North and South America, Europe, Asia, Africa, Australasia

Taking cryptic camouflage to extremes, these three remarkable families include nocturnal insect-eating birds and larger species that also take small frogs and rodents.

Nightjars were called "goatsuckers" because they supposedly take milk from goats. Pliny the Elder, a 1st century Roman naturalist, wrote that nightjars sucked milk from goats, which then became blind. The idea spread even as far as North America. Some European cultures also knew the European Nightjar (*Caprimulgus europaeus*) as the fern owl and may have linked its appearance with dead, unbaptized children doomed to wander in nightjar form. The reality is that the European Nightjar sings a far-carrying churr and is extremely hard to see by day, so it gained a powerful air of mystery when it "appeared" at dusk. In English poetry, it has been called a dor-hawk (after dor beetles), dewfall hawk, or eve-jar (it appears at dusk). It catches large moths in the air, in its wide gape, using a twisting, acrobatic flight. Its tiny legs are unsuitable for walking.

▲ Milk thief
In *Der Naturen Bloeme*, Jacob van Maerlant (c.1280) describes a bird, thought to be a nightjar, that takes milk from a goat.

Vocalizations and ornamentation

Nightjars worldwide have a similar character and sing trills, churrs, or repetitive phrases at dusk. In Africa, the song of the Fiery-necked Nightjar (*C. pectoralis*) is interpreted as "Good Lord deliver us." The Standard-winged Nightjar (*C. longipennis*) has elongated bare shafts with a broad, rounded tip that can be raised above each wing, while the Pennant-winged Nightjar (*C. vexillarius*) has a long, white spike extending behind each wing. South America's Lyre-tailed Nightjar (*Uropsalis lyra*) has a fantastic long, scissor-shaped tail.

> "Flying upon the goat, it sucks them, whence it has its name."
>
> ARISTOTLE, *History of Animals*, c.350BCE

In North America, the Common Nighthawk (*Chordeiles minor*) is more familiar, hunting for moths at dusk over urban areas. The Common Poorwill (*Phalaenoptilus nuttalli*) is the only bird that truly hibernates, becoming torpid for months. It is known to the Hopi people, who call it *holchoko* (the sleeping one).

Bristles around the bill, which help trap food, are characteristic of the flat-headed nightjars and large-headed, rounder-bodied, and more owl-like frogmouths of Southeast Asia and Australasia. The large South American potoos lack bristles and hunt from a perch instead of in flight, more like flycatchers.

◄ Frogmouth family
Viewed head-on, forward-facing eyes give Tawny Frogmouths (*Podargus strigoides*) an owl-like face, but the triangular bill and wide gape are quite different.

▼ Well camouflaged
The patterned plumage of the Great Potoo (*Nyctibius grandis*) provides effective camouflage against a tree.

Order Apodiformes

Families 2

Species More than 110

Size range 3½–12 in (9–31 cm) long

Distribution Worldwide except Antarctica

A treeswift's tail often closes to a point in flight but has a distinctive fork at other times

CRESTED TREESWIFT
Hemiprocne coronata

Very long, very narrow wings give greater lift without using energy, but may reduce maneuverability in tight spaces

A band of white crosses the rump and throat on the Little Swift's otherwise sooty plumage

LITTLE SWIFT
Apus affinis

Long, curving flight feathers give the swift its pointed wingtips

Swifts and Treeswifts

Apodidae, Hemiprocnidae

Soaring through open skies, swifts are a joy to watch. With their long wings and slender bodies, these elegant birds are adapted to an aerial life to the exclusion of all else.

Swifts are easy to spot flying high in clear, open skies, except when actually on the nest. Unlike their relatives the treeswifts, which have a hind toe similar to those of perching birds, these masters of the air are unable to perch on a wire or twig. Instead, their tiny feet have two pairs of toes that point sideways, with sharp, curved claws that allow them to cling to rock faces, walls, or sandy banks. When a swift accidentally lands on the ground, it struggles to rise again, due to its short legs and long wings.

At home in the sky

With long, tapered wings that exceed their body length, swifts are supremely adapted to a life in the air. They are able to glide for hours without wasting precious energy, and can eat, mate, gather nesting material, and even sleep in the air. Despite their name, these birds often fly rather slowly when foraging, but are capable of rapid bursts of speed, especially during communal display flights. The large needletail swifts

▼ Taking a sip
Like other swifts, the Common Swift (*Apus apus*) dips to the surface of a lake or stream when it needs to drink.

Like other swifts and treeswifts, palm swifts have a short bill with a very wide gape to trap flying insects

◄ Aerial acrobats

The crested treeswift of southeast Asia is a woodland specialist that builds a cuplike nest on an open branch. The little swift, from Africa and India, is square-tailed and broad-winged. The African palm swift, like most swifts, eats insects and airborne spiders but builds its nest among the fronds of a palm tree.

AFRICAN PALM SWIFT
Cypsiurus parvus

The common swift spends **two years** in flight before nesting

Palm swifts tuck their tiny legs away in flight to reduce drag

" Fifteenth of May. Cherry blossom. The swifts
Materialize at the tip of a long scream
Of needle. 'Look! They're back! Look!' . . . "

TED HUGHES, "Swifts" 1974

(*Hirundapus* spp.) of Asia are the fastest birds recorded in level flight, reaching speeds of around 105 mph (170 kph). Cave-nesting swiftlets provide the key ingredient for bird's-nest soup, a highly prized dish that has been consumed in China for hundreds of years. Today, special factories have been built of sand and clay that mimic the swiftlets' caves. Wild swiftlets nest in these multistory buildings, and the nests are then harvested by workers and sold for vast sums.

► Clinging to the rockface

In Indonesia, the Cave Swiftlet (*Collocalia linchi*) sticks its edible, cuplike nest of plant matter to the rock face with saliva. These swiflets use echolocation when navigating deeper inside the dark caves.

Order Apodiformes

Families 1

Species More than 350

Size range 2¼–4⅔ in (5.7–12 cm) long

Distribution North and South America

◄ A charm of hummingbirds
Biologist Ernst Haeckel's whimsical painting from 1899 recalls the glass case collections of the time. In these displays, a variety of hummingbird specimens and other species, which would never be found together in the wild, were put on show.

The thin bill and long tongue are inserted into tubular blossoms to feed

The male's tail is purple, whereas the female's feathers have a black band and white tip

◄ Shimmering feathers
The Green-breasted Mango (*Anthracothorax prevostii*) of Central America glistens with iridescent green, blue, and purple, but its wings are dull.

Hummingbirds

Trochilidae

Hummingbirds are small but remarkable. They display spectacular iridescent colors, have extraordinary aerobatic abilities, and can even fly backward.

Despite being restricted to the Americas, hummingbirds are among the most popular birds worldwide. These famously small, high-energy birds have a wingbeat speed so rapid that the human eye cannot comprehend them. Specially adapted sockets at the base of the wing allow it to twist and create lift on the forward and backward strokes, thus enabling the bird to fly backward.

Long-billed species such as the Sword-billed Hummingbird (*Ensifera ensifera*) feed on tubed flowers, while others, including the Fiery-throated Hummingbird (*Panterpe insignis*), pierce flowers at the base with their short bills. Some species, such as the Tooth-billed Hummingbird (*Androdon aequatorialis*), saw into a bloom. These methods are used to access nectar to fuel the bird's fast metabolism. Many species also feed on small insects.

Hummingbirds have nondescript songs—their name comes from the noise their rapidly beating wings make as they fly—but some perform

Only the adult male has the stunning blue and purple "headdress" and mask

► Standing out
Costa's Hummingbird (*Calypte costae*) breeds in southwestern US and Mexico, adding tiny flashes of color to the arid chaparral and sage scrub.

BEE HUMMINGBIRD
Mellisuga helenae

Restricted to the forests of Cuba, this is the smallest bird in the world at just 2¼ in (5.7 cm) long. It eats more than its own bodyweight—1⁄16 oz (2 g)—in nectar each day.

FIERY-THROATED HUMMINGBIRD
Panterpe insignis

Found only in Costa Rica and western Panama, this species occupies high-altitude cloud forest and dwarf forest, up to 10,500 ft (3,200 m) high, and open woods at lower altitudes.

RUFOUS-CRESTED COQUETTE
Lophornis delattrei

Found near the Pacific coast of South America in clearings and forest edges, this distinctively crested bird pumps its tail during flight, which creates a beelike motion.

A hummingbird is able to **lap up nectar** at a rate of about **14 licks per second**

The flexible tail feather shafts cross and curve out to the opposite side

The tips of the male's elongated tail feathers rise and sway in flight

"... the little bird darted its beak into the wild flowers, making an extraordinary buzzing noise ... with its wings."

CHARLES DARWIN, *The Voyage of the Beagle,* 1839

remarkable displays. Their colors are unusual, because many of them are created by iridescence, in which the feather surface has two minute layers and light reflects back from both in slightly unequal wavelengths, creating a vibrant effect. These bright patches depend on the angle of view, and can be turned toward a rival or potential mate like a flashlight in the dark vegetation.

Legendary bird

Hummingbirds appear in an abundance of myths and legends of Indigenous North American and South American peoples. They are associated with rain, and many tales relate to a loss that was later compensated for by the brilliance of their plumage. Pueblo peoples related how a demon lost a bet with the Sun and decided to set fire to the Earth. A hummingbird gathered clouds that put out the fire with rain; in doing so, the bird flew through rainbows and took on their vivid colors.

The Hitchiti people told of a conflict over food between the heron and the hummingbird. The birds raced to a treetop to settle their differences, but the hummingbird lost because it fed on flowers along the way. Since then, the

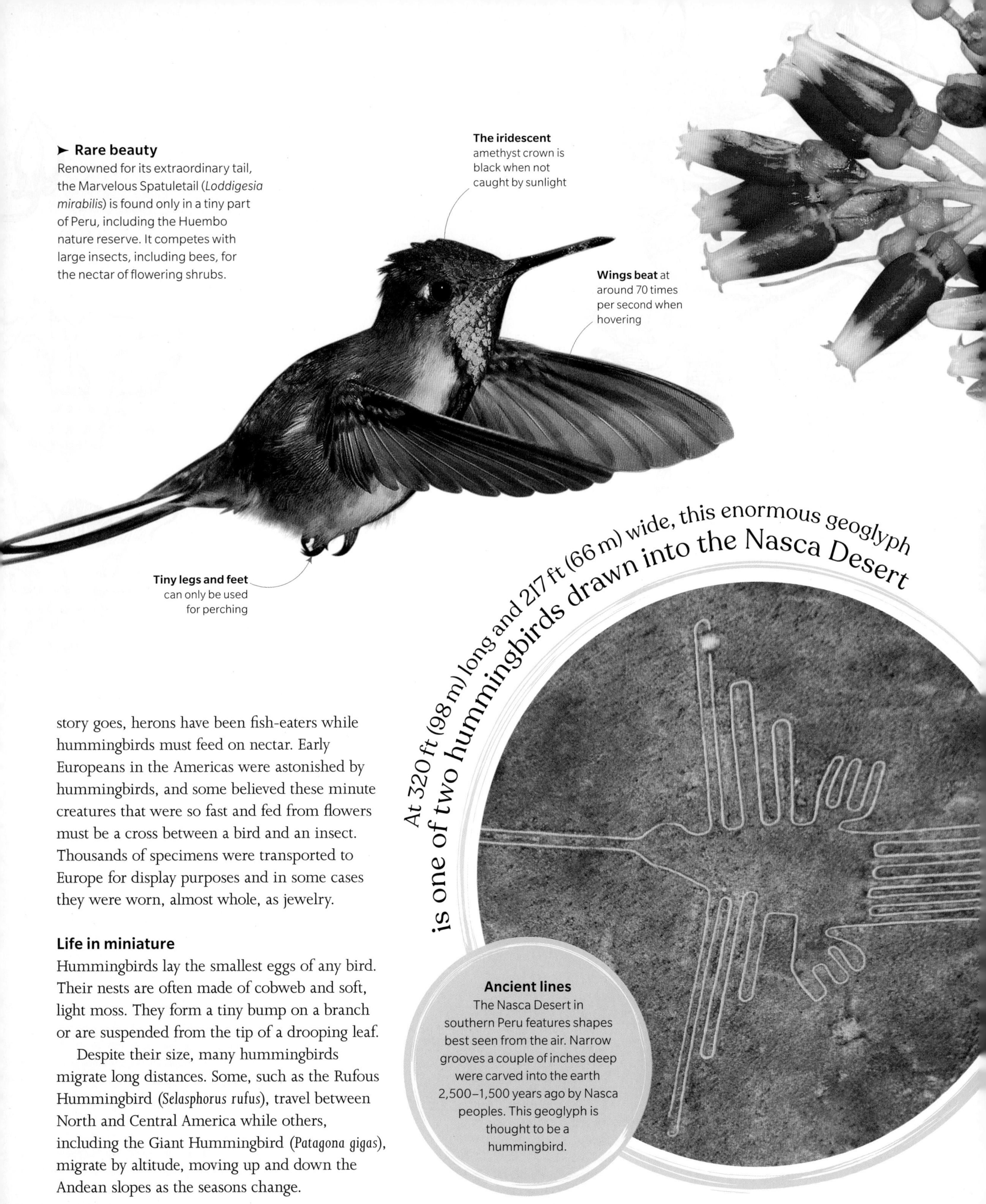

➤ Rare beauty
Renowned for its extraordinary tail, the Marvelous Spatuletail (*Loddigesia mirabilis*) is found only in a tiny part of Peru, including the Huembo nature reserve. It competes with large insects, including bees, for the nectar of flowering shrubs.

Ancient lines
The Nasca Desert in southern Peru features shapes best seen from the air. Narrow grooves a couple of inches deep were carved into the earth 2,500–1,500 years ago by Nasca peoples. This geoglyph is thought to be a hummingbird.

story goes, herons have been fish-eaters while hummingbirds must feed on nectar. Early Europeans in the Americas were astonished by hummingbirds, and some believed these minute creatures that were so fast and fed from flowers must be a cross between a bird and an insect. Thousands of specimens were transported to Europe for display purposes and in some cases they were worn, almost whole, as jewelry.

Life in miniature

Hummingbirds lay the smallest eggs of any bird. Their nests are often made of cobweb and soft, light moss. They form a tiny bump on a branch or are suspended from the tip of a drooping leaf.

Despite their size, many hummingbirds migrate long distances. Some, such as the Rufous Hummingbird (*Selasphorus rufus*), travel between North and Central America while others, including the Giant Hummingbird (*Patagona gigas*), migrate by altitude, moving up and down the Andean slopes as the seasons change.

Proceeding with caution
A somewhat stylized common moorhen cautiously makes its way through a marsh in this print by Japanese artist Ohara Koson.

▲ Bold and beautiful
The Western Swamphen (*Porphyrio porphyrio*) is found in southern Europe and northwestern Africa. It defends its territory with aggressive displays and fights.

Rails, Crakes, and Coots

Rallidae

Although members can be found in various habitats around the world, most of this family can be split into reedbed-loving crakes, waterside-dwelling rails, and the more aquatic coots.

Rails, crakes, and coots form a diverse group of birds that usually live near water. They have some common features, such as short wings, long toes, and pointed bills, but they also vary in size, color, and behavior. They are not prominent in many stories or legends, but they have given rise to some popular sayings. For example, "bald as a coot" refers to the large, bare, fleshy frontal shield on the forehead of coots (usually white) and moorhens (often red). "Crazy as a coot" may relate to the bird's sudden "escape" when disturbed on dry land. Its instinct is to run for water, wings flapping, making loud, unmusical calls.

Most rails, crakes, and coots are waterside rather than strictly water birds. Coots (*Fulica* spp.), however, spend more time on open water, often in large flocks, where they dive for food that they bring up to the surface. They form

Order Gruiformes

Families 1

Species More than 150

Size range 5–25 in (12–63 cm) long

Distribution Worldwide except Antarctica

▲ Kick boxing
Two Eurasian Coots (*Fulica atra*) fight for territorial dominance, revealing the lobed toes that separate coots from rails. These help coots to swim and dive in open water; they behave more like diving ducks or grebes.

flocks with other species, including surface-feeding ducks, such as gadwalls, which gather to steal discarded waterweeds and any small prey that may be revealed, while gulls swoop overhead.

Rails and crakes are usually much more elusive. They prefer the interior of reedbeds or fen vegetation, often appearing just when feeding on a patch of bare mud. They are more often heard than seen, especially species such as North America's Clapper Rail (*Rallus crepitans*), whose clattering duets and choruses ring around the marsh, or Eurasia's Water Rail (*Rallus aquaticus*) with its piglike squeals. The Common Moorhen (*Gallinula chloropus*) calls repeatedly at night during flights around its territory.

Some species of rail and crake have very limited ranges. For example, the Invisible Rail (*Habroptila wallacii*) is endemic to Halmahera island in Indonesia, while the Inaccessible Island Rail (*Laterallus rogersi*) is found only on a remote island in the South Atlantic. No bigger than a newly hatched farmyard chick, it is the world's smallest flightless bird and numbers fewer than 12,000 individuals. At least 26 species of rails have become extinct in historical times, mainly because they had limited ranges and small populations and were vulnerable to introduced predators, such as cats or rats. The Galápagos Crake (*L. spilonota*), found on seven of the Galápagos Islands, is beset by pressures from overgrazing by cattle and goats, predation by rats and cats, as well as the effects of climate change.

Back from the brink

One large flightless rail, the Takahē (*Porphyrio hochstetteri*) from New Zealand, saw its numbers fall below 200 in the 1980s. This species is valued by the Ngāi Tahu tribe, some of whom have worked together with conservation organizations to help increase numbers. A dry grassland species, the takahē faces threats from introduced species, such as stoats and deer. A captive-breeding program began in 1985, and the release of birds onto predator-free islands has helped numbers increase to around 500.

In much of Eurasia, the Corn Crake (*Crex crex*), with its harsh, two-note song, provides a familiar sound in fields and meadows in early summer.

➤ New Zealand giant
The takahē is the world's largest living rail, measuring 25 in (63 cm) long. This 19th-century illustration of a breeding pair by Elizabeth and John Gould and Henry Richter may have been copied from a museum exhibit rather than live birds.

CORN CRAKE
Crex crex
A dry-land bird that migrates between Africa and Eurasia, this crake is adapted to rough grass with clumps of coarser plants. It feeds on insects, spiders, slugs, snails, worms, and seeds.

SORA
Porzana carolina
Found only in the Americas, the sora can be distinguished from other rails and crakes by its short yellow bill and black face and throat. It is a migrant between North and South America.

OKINAWA RAIL
Hypotaenidia okinawae
Sporting a bright red bill and legs, this rail is only found on the Japanese island of Okinawa. It is almost flightless and has learned to roost in trees to avoid predation by snakes.

Despite its short wings, trailing legs, and a weak-looking, heavy flight, this bird manages an annual migration from Europe and western Asia to southern Africa. In Russia and Kazakhstan, its numbers are still substantial, but in western Europe, its population has declined drastically due to changes in farming practices, such as earlier harvesting of crops and the introduction of the mechanical cutting of hay in the 20th century. These have made it difficult for pairs to rear chicks. Once widespread, corn crakes are now restricted to remote areas, such as islands off Ireland and Scotland. Payments to farmers to encourage later harvesting have seen some success, but the species still needs a lot of support from local conservationists.

➤ Elusive marsh bird
This water rail shows the long bill, striped back, and barred flank common to more than a dozen rail species worldwide.

Long legs and a compressed, slender body suit creeping through dense reedbeds

Order name Gruiformes

Families 1

Species 15

Size range 35–70 in (75–90 cm) tall

Distribution North America, Europe, Asia, Africa, Australasia

▲ **Trailing legs**
Cranes, such as this Common Crane (*Grus grus*), fly with their long legs and large feet trailing, as these may simply be too large to be tucked forward into their body feathers.

Cranes

Gruidae

Elegant, upstanding, and noisily vocal, cranes are famous for their elaborate courtship rituals. Those that succeed in forming a breeding pair usually remain together for life.

Cranes are large, long-legged, and long-necked—the Sarus Crane (*Antigone antigone*) is the world's tallest flying bird—but not as long-billed as a heron or stork. They have a long-striding walk over open ground and marshland, seeking a variety of food from small rodents and fish to berries and grain. They may feed on drier ground, but return to the safety of wetlands at night. Social birds, they gather in flocks outside the breeding season and prebreeding groups are especially exciting, when pairs indulge in elegant leaping and dancing displays. Unlike storks and herons, they are not colonial nor do they nest in trees; they make large nests on wet ground or in shallow water. Their calls are individually recognizable, even though the voice may change with age, and scientists can

The call of a common crane can be heard from more than **3 miles (5 km) away**

> " A falcon who chases a warlike crane can only hope for a life of pain. "
>
> GABRIEL GARCIA MARQUEZ, *Chronicle of a Death Foretold*, 1981

Cranes fly with their long neck extended, like storks but unlike herons, which hunch the head back

Narrow wingtips are fingered and curve upward under pressure, reducing turbulence

distinguish pairs from year to year by their voices. Families remain together over winter and most crane species migrate long distances to wintering areas in flocks.

Symbols of love and virtue

Most breeding pairs will remain together for life and cranes have become symbols of love and fidelity, although a small percentage do "divorce," especially if breeding attempts fail in younger pairs. In the 1st century CE, Roman naturalist Pliny the Elder wrote that cranes appointed a vigilant sentry while others slept. That individual would hold a stone in one raised foot, so that if it should fall asleep, the dropped stone would awaken it. In many cultures, especially in Asia, cranes are associated with happiness and a long life or, alternatively, with eternal youth.

Several species have been the subject of conservation efforts. The Whooping Crane (*Grus americana*) was reduced to just 23 individuals (two captive) by 1941. Protection along its migration routes from Canada to the Gulf of Mexico saw numbers recover and breeding pairs have been reintroduced into the wild. The Siberian Crane (*Leucogeranus leucogeranus*) numbers around 3,500 birds. At the other extreme, North American Sandhill Cranes (*Antigone canadensis*) winter in flocks of up 10,000 and some areas see close to half a million on migration each year.

Divine birds
This scroll painting from the 1700s by Chinese artist Chen Quan shows Red-crowned Cranes (*Grus japonensis*). Despite being among the rarest of cranes, they are associated with immortality in Taoist legends. In Japan, they are sometimes called "the Gods of the marshes."

◄ Mudflat roost
Most members of this family are gregarious and often feed together. Here, a large flock of Pied Avocets (*Recurvirostra avosetta*) have gathered to roost in Lianyungang, China.

Stilts and Avocets

Recurvirostridae

Stilts and avocets are renowned for their exceptionally long legs, which enable them to wade in waters too deep for most other shorebirds while foraging for food.

Order Charadriiformes

Families 1

Species 10

Size range 14–20 in (35–51 cm)

Distribution
Worldwide except Antarctica

Stilts have long pink legs, while avocets have long blue-gray legs; both groups have a long neck and a long, very narrow bill that is straight in stilts and upwardly curved in avocets. Apart from the Black Stilt (*Himantopus novaezelandiae*), all members of the family show bold patterns of black and white, and some sport brown, orange, or pink patches.

Feeding and breeding

Their long legs allow stilts and avocets to wade in deep water and feed on prey out of reach of other shorebirds. Both groups swim, but avocets have partially webbed toes so they are more efficient. While avocets mainly feed on large numbers of crustaceans, stilts have a broader diet that includes insects, tadpoles, worms, fish, and crustaceans. Avocets often "scythe" for food, moving their bill from side to side to touch-detect prey. Stilts can scythe too, but more often they dart down and quickly grab prey spotted in the water. The avocets and Australia's Banded Stilt (*Cladorhynchus leucocephalus*) have mandibles with lamellae—rows of hairlike structures that mesh together and filter out inedible or large items—and an enlarged tongue that pushes trapped food toward the mouth.

Stilts and avocets are very noisy, especially around the nest. Unusually for shorebirds, they nest in colonies. These are usually of modest size, 10–70 pairs, but the banded stilt breeds in much greater numbers. The birds wait for rain to fill huge inland salt lakes before settling and feeding on the lakes' abundant brine shrimp. Such colonies have been known to reach 179,000 pairs.

▲ RSPB logo
The pied avocet is the logo of the UK's Royal Society for the Protection of Birds, celebrating the successful conservation of the species since its return in the 1940s.

The **banded stilt** can nest at very high densities of up to **18 nests** per 10 square feet

BLACK STILT
Himantopus novaezelandiae

The Black Stilt is a critically endangered bird from New Zealand. In 1981, there were only 23 in the wild, but conservation efforts have raised their number to about 100.

AMERICAN AVOCET
Recurvirostra americana

American Avocets specialize in breeding on ephemeral wetlands in the US Midwest. Like others in the family, they are highly aggressive and noisy in defending their colonies from predators.

▲ Mating ritual
After mating, the male black-winged stilt often places a wing across the female as the two cross bills and walk together

Plovers and Lapwings

Charadriidae

▲ The quad squad
Most plovers lay three or four eggs. This 1873 illustration by John Gould shows a pair of Northern Lapwings (*V. vanellus*), whose eggs have successfully hatched.

This family of small- to medium-sized shorebirds is adapted to finding prey by sight in open spaces. Known for their loud calls and showy display flights, these graceful birds are found all over the world.

Order Charadriiformes

Families 19

Species More than 60

Size range 5½–15 in (14–38 cm)

Distribution
Worldwide except Antarctica

At first glance, plovers resemble other shorebirds, such as the sandpipers that share their estuarine, coastal, and tundra habitats. However, the short bill and large head and eyes, and distinctive feeding method set them apart. Those large eyes give a clue as to what they are doing: unlike most sandpipers, plovers are sight feeders, rarely using touch for foraging, and seldom probing mud or soil for prey with their bill. They have excellent eyesight, and their eyes have a high density of rod cells (photoreceptor cells in the retina), enabling

◄ Hidden treasures
Three Black-bellied Plover chicks (*P. squatarola*) huddle in their nest. Plover nests consist of little more than a scrape on the ground.

The white patch on the nape of these chicks is amazingly effective at breaking up their shape when seen from above

them to hunt effectively even in low light conditions. They have a distinctive "stop-run-peck" feeding method—they stand still, surveying their immediate surroundings. If nothing is visible, they make a short run in any direction and stand still again. As soon as they spot something, they run toward it and stoop down to grab it. This method enables plovers to catch all kinds of invertebrates, from large worms on estuaries to midges on the Arctic tundra. They also sometimes eat berries.

There are two main groups in the plover family. The "ringed" plovers tend to be small and sharp-winged, and many have brown or black chest bands (incomplete in species living in habitats made up of a uniform color, such as sandy beaches, and complete in species living in habitats with a variable background, such as gravel). Lapwings (*Vanellus* spp.) are larger and taller, bolder in color and often with head adornments such as crests or wattles. Some species have rounded wings.

High flyers

Some plovers travel great distances: the American Golden Plover (*Pluvialis dominica*) makes one of the world's longest landbird migrations, flying up to 12,500 miles (20,000 km) from northern Alaska and Canada to Argentina. This includes a single leg of almost 2,500 miles (4,000 km) over the Atlantic Ocean. The Pacific Golden Plover (*P. fulva*), meanwhile, can reach an altitude of more than 2.5 miles (4 km) during its migration across the Pacific Ocean.

Many species perform showy display flights on the breeding grounds, and these are often accompanied by loud calls. They also have various distraction displays, such as the "broken wing" ruse to distract predators from their chicks. Some can be extremely aggressive toward potential threats to the nest, often swooping at people. In South America, the aggressive behavior of the Southern Lapwing (*V. chilensis*) earned it the respect of ranchers, who appreciated that it warned livestock of potential predators. Farmers tethered these birds or clipped their wings, forcing them to stay on their land and act as a kind of "guard dog."

RED-WATTLED LAPWING
Vanellus indicus

Common all over India and Southeast Asia, this is a noisy and familiar bird. There are many local folk-names that mimic the bird's call—one is "Hatatut."

KENTISH PLOVER
Charadrius alexandrinus

These birds live on sandy beaches on the coast, and also on saline or brackish wetlands far inland. Adults sometimes partially bury their eggs under sand during the heat of the day.

Most plovers have long wings with sharp wingtips, enabling them to fly fast and far

This is a male dotterel; the females have a cleaner cut chest band

► Wings aloft
The Eurasian Dotterel (*Charadrius morinellus*) breeds in the Arctic tundra. The male undertakes most parental duties, including incubation and brooding.

➤ White chancer
A small percentage of male ruffs are mainly white and do not display at a specific lek. These "satellite" males move between leks and mate opportunistically.

◄ Wheeling flocks
Typical of the family, bar-tailed godwits gather in large, tight flocks on estuaries and mudflats outside the breeding season. A long, narrow bill is ideal for probing for worms and shellfish.

Sandpipers and Snipes

Scolopacidae

Order Charadriiformes

Families 1

Species More than 90

Size range 4¾–26 in (12–66 cm) long

Distribution Worldwide except Antarctica

Known for their epic migrations, swirling flocks, remarkable variety of bill shapes, and odd breeding antics, the members of this family have long captured the imaginations of people worldwide.

The sandpipers and snipes are commonly found at estuaries, mudflats, beaches, and lake shallows around the world—and also on tundra, marshes, and bogs during the breeding season. Known as shorebirds or waders, they are small to medium-sized running and walking birds, often with cryptically colored plumage that serves as camouflage. Their long wings are narrow and pointed, enabling them to fly at high speed. Family members exhibit an extraordinary range of bill shapes, related to the food they consume and how they do so. The sparrow-sized stints (*Calidris* spp.) have a short, straight bill and pick small food items from the surface, spotted by sight. Tall, gull-sized curlews (*Numenius* spp.) have a very long, down-curved bill, ideal for probing inside the burrows of crabs and worms, or in soft mud, mainly using touch. The Ruddy Turnstone (*Arenaria interpres*) uses its short bill to lift stones and weed to reveal prey. Touch feeders have vast

▲ **Brightest feathers**
The Eurasian Woodcock (*Scolopax rusticola*) has the world's brightest feathers on its tail tips, which are displayed in courtship flights at dusk.

concentrations of chemical and touch receptors at the bill tip, allowing them to evaluate feeding sites and find food efficiently. With eyes on the side of the head, sandpipers and snipes can scan panoramically for predators.

Feeding and breeding

The majority of family members breed at high latitudes on tundra or bogs and migrate south after breeding. Several species, such as ruddy turnstones and Sanderlings (*Calidris alba*), seem to disperse to every corner of the world, including remote islands. Bar-tailed Godwits (*Limosa lapponica*) from Alaska undertake a nonstop journey in fall to spend the winter in Australia or New Zealand. In 2022, one tracked individual flew 8,425 miles (13,560 km) in 11 days and 1 hour, without touching down. Once on the wintering grounds, many species are drawn to intertidal estuaries. They feed according to the tide, and when it is high they head to roosts, often swirling around in large flocks making spectacular aerial maneuvers.

On their breeding grounds, many species perform eye-catching display flights in courtship, often with a rich vocal accompaniment. The Common Snipe (*Gallinago gallinago*) is known for its "drumming" display, in which the outer tail feathers vibrate to make a bleating sound. Sandpipers and snipes have an extraordinary variety of mating systems. Female Spotted Sandpipers (*Actitis macularius*) may lay four clutches, each incubated by a different male. Ruffs (*Calidris pugnax*), Great Snipes (*G. media*), and Buff-breasted Sandpipers (*C. subruficollis*) all perform courtship rituals at leks. Unusually, ruffs display in complete silence, which may explain the extraordinary individual coloring of males.

Snipes insert their bill into sediment to locate worms and other food by touch

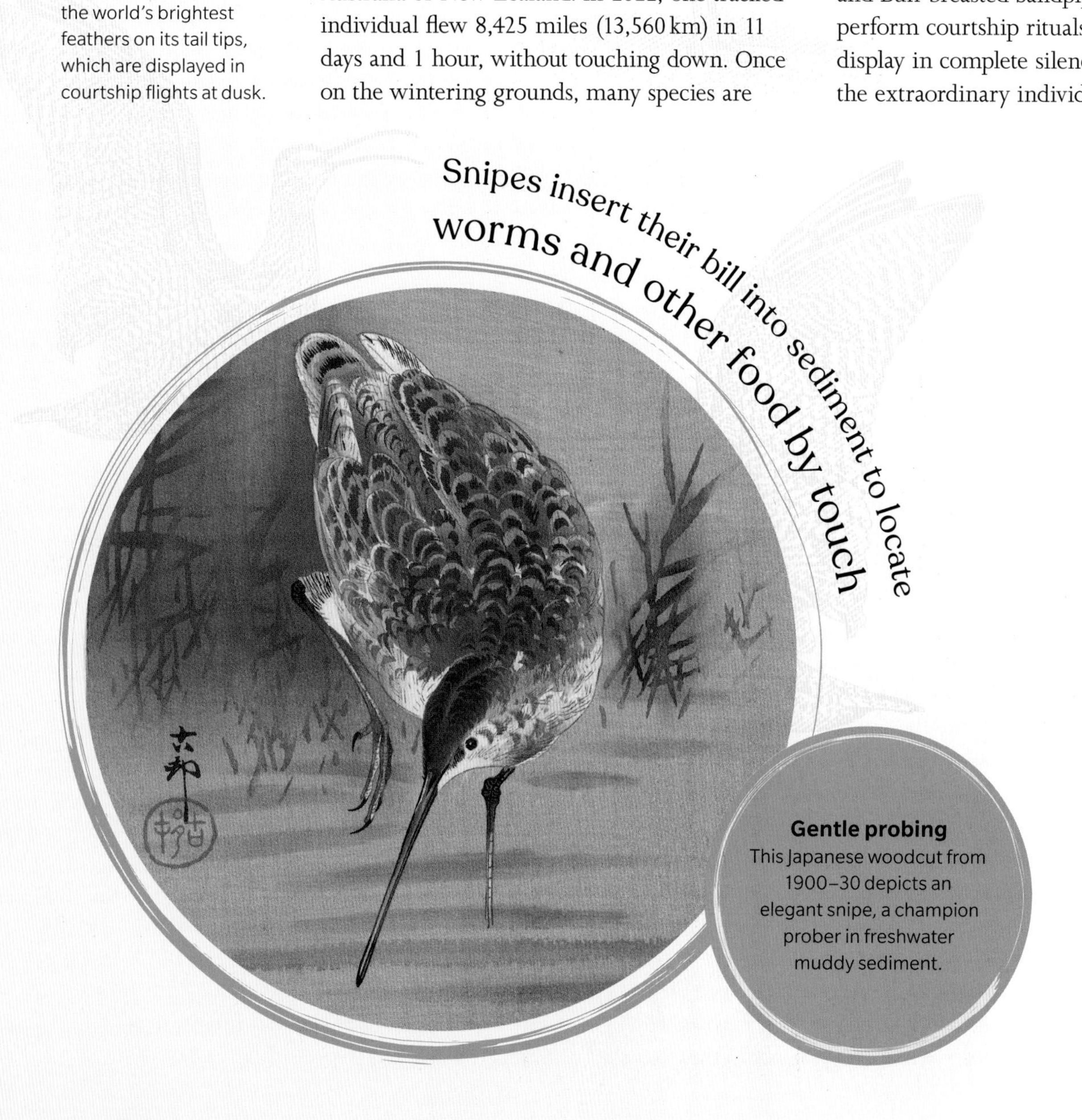

Gentle probing
This Japanese woodcut from 1900–30 depicts an elegant snipe, a champion prober in freshwater muddy sediment.

PECTORAL SANDPIPER
Calidris melanotos
Males of this tundra-breeding species have an inflatable air sac on their chest and make bizarre hollow hooting sounds during courtship flights.

RED PHALAROPE
Phalaropus fulicarius
Unlike other waders, phalaropes have lobed toes, helping them to be strong swimmers, and they spend the winter far out at sea.

Order Charadriiformes

Families 1

Species 8

Size range 6–20 in (15–50 cm) long

Distribution North and South America, Africa, Asia, Australia

➤ **Floating home**
Also known as the lily trotter, the African Jacana (*Actophilornis africanus*) lives largely on unstable masses of floating plants, especially waterlily pads.

Jacanas

Jacanidae

These waterbirds have greatly elongated toes, allowing them to walk over floating vegetation.

Found throughout warmer parts of the world, jacanas are adapted to an unusual habitat: floating and emergent vegetation growing over freshwater lakes and marshes. Their long, thin toes are splayed so broadly that they can walk or run where other waterbirds might sink. They have a relatively long bill, which is used to glean prey, such as insects and crabs, and also to turn over leaves to reveal hidden prey. Some jacanas also eat vegetation, including waterlily flowers.

Role reversal

Apart from the monogamous Lesser Jacana (*Microparra capensis*), jacanas have unusual breeding arrangements. The males take on all the nest-building, incubation, and chick-rearing roles, while the females, which may be almost twice as heavy, may on occasion help guard the family from predators. Both males and females defend territories against same-sex rivals, and it is common for females to mate with several males in the same season, sometimes as many as four. The males build many floating platforms, including for solicitation purposes, and nests are formed from a "cup" of waterweed placed on the unsteady surface. In the shallows, nest losses are high, but males can move eggs to a new nest and also protect and carry chicks under their wings. Occasionally, a female will intrude upon a brooding male and kill his previous partner's young, inducing the male to mate again.

▼ **Big feet**
This c.1845–49 illustration shows a Northern Jacana (*Jacana spinosa*), whose toes can cover an area of 4 ¾ x 5 ½ in (12 x 14 cm).

Toes and claws are splayed to distribute weight

The hooked bill is an effective weapon, used for hunting lemmings and fish, or snatching prey from other birds

Order Charadriiformes

Families 1

Species 7

Size range 15–23 in (38–58 cm) long

Distribution Worldwide

GREAT SKUA
Stercorarius skua
Able to knock a gannet into the water, this powerful skua sometimes kills or robs other seabirds, but most of the time it catches fish as it roams the Atlantic Ocean.

SOUTH POLAR SKUA
Stercorarius maccormicki
Often seen at penguin colonies on the lookout for injured chicks, this large skua breeds on the Antarctic coast but makes an annual migration to the Northern Hemisphere.

▲ Elegant hunter
Long-tailed jaegers are usually pale, with an unmarked white breast becoming darker toward the tail. As with other jaegers, all-dark forms exist too, but they are very rare.

A breeding adult has striking tail streamers that are around 6 in (15 cm) long; they are shed for the rest of the year

Long, slender wings allow the jaeger to travel vast distances at sea—it only comes to land to nest

➤ Varying plumage
This 1910 watercolor by renowned US ornithologist Louis Agassiz Fuertes depicts an adult pomarine jaeger (top), dark, juvenile, and pale parasitic jaegers (right), adult and juvenile long-tailed jaegers (bottom left), and a great skua (bottom right).

Skuas and Jaegers

Stercorariidae

Renowned for their audacious midair heists, snatching food from seabirds of all sizes, skuas and jaegers are often described as the pirates of the skies.

With long wings and a short, hooked bill, skuas and jaegers are fearsome predators that can perform amazing aerial maneuvers when hunting or stealing food from other seabirds. Skuas are larger with a short tail, while the jaegers are smaller with a tail that extends into a central spike or two streamers. This spike is twisted and spoonlike on the Pomarine Jaeger (*Stercorarius pomarinus*), but it is often broken—or even intentionally bitten off—once spring courtship displays are complete.

All skuas and jaegers undertake long-distance migrations over the oceans, and these birds are only seldom seen inland beyond their breeding grounds, unless they are driven there by storms. Southern-Hemisphere skuas migrate as far north as Alaska, while Arctic-breeding jaegers fly as far south as New Zealand, covering as much as 190 miles (300 km) in a day.

Approach with caution

On breeding territories, jaegers perform dramatic display flights accompanied by noisy calls. Skuas attack anyone approaching a nest in a painful head-height swoop (vividly described by members of Captain Scott's 1911–12 expedition to the South Pole), and Long-tailed Jaegers (*Stercorarius longicaudus*) will even settle on an intruder's head. Northern species have suffered in recent outbreaks of bird flu, but usually their numbers are regulated by their prey, be it lemmings in the Arctic or the seabirds from which they steal food.

The jaeger's large wings give it great maneuverability

▲ Preparing to attack
Jaegers and skuas carry out kleptoparasitism, forcing other birds to give up their hard-caught fish. This juvenile Parasitic Jaeger (*Stercorarius parasiticus*) is targeting a Sandwich Tern (*Thalasseus sandvicensis*).

Fancy feathers and more
Some auks sport striking ornamentations in the breeding season, to appeal to potential mates and also intimidate rivals. The Crested Auklet (*Aethia cristatella*) grows a magnificent quiff and bright orange bill-plates, but also develops a strong, fruity odor.

All three species of puffins shed the colorful plates that cover **their bill** during their annual molt

Auks and Puffins

Alcidae

Order Charadriiformes
Families 1
Species More than 20
Size range 6–18 in (15–45 cm) long
Distribution Northern Hemisphere coasts and seas

These deep-diving seabirds are powered by their wings rather than their feet when swimming underwater. They are plump yet streamlined with dapper black-and-white plumage, and on land have a waddling walk.

Auks are not related to penguins but the first bird ever known as a "penguin" was the now extinct Great Auk (*Pinguinus impennis*). When Europeans first discovered the true penguins, they named them after this similar-looking flightless bird. Unlike penguins, all auk species living today can fly. Taking to the air with their short, flipperlike wings is hard work—the Thick-billed Murre (*Uria lomvia*), a heavy-bodied bird able to dive down to 690 ft (210 m), has the most energy-expensive flight of any animal. However, flying enables auks to breed in safe, inaccessible places such as steep cliff faces or even (in the case of some murrelets) high in rainforest trees, 3 miles (5 km) or more from the coast.

Taken for food

The people of Hirta in St. Kilda, an archipelago off the coast of Scotland, UK, relied on seabirds, including Razorbills (*Alca torda*), Common Murres (*U. aalge*), and Atlantic Puffins (*Fratercula arctica*), as food and risked terrible accidents to reach

> "I wear make-up. . . I apply it for the simple reason that I want to look like [the] charismatic seabird the puffin."

CAITLIN MORAN, *More Than a Woman*, 2020

Thick, dense plumage traps air and helps keep the puffin warm when it dives deep in cold sea water

The colorful plates that the Atlantic puffin grows on its bill exhibit photoluminescence under ultraviolet light, which is visible to other puffins

The stretchy yellow skin around the bill base helps accommodate lots of fish, when the puffin is hunting

➤ Tears of a clown
The colorful bill and melancholy expression of the Atlantic puffin give it a unique and appealing look, almost universally recognizable to people living in the countries where it occurs.

➤ Cliff colony
Thick-billed murre colonies hold thousands of pairs, each defending a tiny territory around its nest.

TUFTED PUFFIN
Fratercula cirrhata

The largest puffin, this northern Pacific species has a huge orange bill, scarlet feet, and creamy-yellow "eyebrows" during the breeding season.

PARAKEET AUKLET
Aethia psittacula

Parakeet Auklets are highly vocal on their breeding grounds, producing various loud trilling and grunting notes when reunited with their mate.

COMMON MURRE
Uria aalge

This large, sleek Atlantic auk is noted for its pointed, almost conical eggs, the shape helping them to sit more stably on the narrow nesting ledges.

their nests. Life was tough, and the last people left for the mainland in 1930. Today, the uninhabited archipelago is a nature reserve and UNESCO World Heritage Site, home to a million auks and other seabirds. Visitors can still explore Hirta's small drystone buildings in which harvested seabirds were dried and stored for winter food (the auks only come to land to breed between April and August, spending their winters far offshore).

Threatened or thriving

Today, auks are threatened by sea pollution, depleted fish stocks, the impact of climate change, and in recent years by outbreaks of avian flu. Some species, such as the Guadalupe Murrelet (*Synthliboramphus hypoleucus*), are classified as endangered. However, vast colonies of Dovekies (*Alle alle*) and common murres still thrive on remote Arctic coasts and islands.

All auks dive to catch prey, primarily fish but also squid and other marine invertebrates. They are much more comfortable on water than land—their short legs, set near the rear of the body, and webbed feet are great for swimming on the surface, but make them clumsy walkers.

Accordingly, many species have a very short nestling stage and "fledge" well before they can fly, heading for the safety of the sea at just a couple of days old in the case of the Ancient Murrelet (*S. antiquus*). Encouraged by a parent, the chicks scramble or even leap headlong from their nest, hoping for a safe landing on the water below.

➤ Captured and collared
The flightless great auk nested on Arctic islands until 1844. Sadly, it was hunted to extinction for its feathers and body fat, and to provide museum specimens. This illustration from 1655 shows a collar around the neck, as this bird was kept as a pet.

▲ Breeding suit
At the end of the breeding season, the Black Guillemot (*Cepphus grylle*) of the North Atlantic will molt from the blackish breeding plumage seen here into almost pure white winter attire.

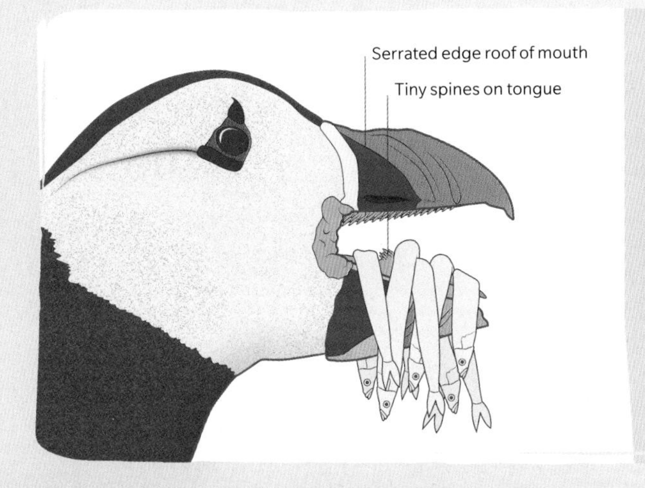

KEEPING HOLD OF A CATCH

Most auks will capture one fish at a time, but a puffin can line up and secure several small fish, such as sandeels, in its bill. It is able to do this thanks to inward serrations on its bill's edges, spines on the roof of its mouth, and a strong, raspy tongue. It brings its bounty back to its nest, flying fast and then running quickly into its burrow to dodge gulls and skuas, which try to steal its hard-earned catch from it.

PUFFIN WITH A BEAKFUL OF FISH

Order Charadriiformes

Families 1

Species More than 100

Size range 8–27½ in (20–70 cm) long

Distribution Worldwide

▲ Unique beak
When the Black Skimmer's (*Rynchops niger*) lower mandible strikes a fish, its bill snaps shut with a backward jerk of the head, thereby trapping the prey.

Gulls, Terns, and Skimmers

Laridae

This mixed group of water and waterside birds lives around fresh water and open ocean, yet some family members are also familiar sights in towns and on farmland.

Unlike most birds, the black skimmer has **vertical pupils**, which narrow to slits to **reduce the glare of water** and sand

In general, gulls are larger than terns. Most of them have longer legs and are able to walk around more freely than short-legged terns, which do not move far on land. Kittiwakes (marine gulls) are the exception. Adapted to nest on narrow cliff ledges, they have short legs, with a sloping-backed stance, and are more suited to a life at sea than foraging ashore. Skimmers—three very similar black and white species—have a ternlike appearance, but show one remarkable adaptation. In most bird species, the upper and lower mandibles are of equal length, or sometimes the upper one is longer. Skimmers have the lower half of the bill protruding. This distinctive feature restricts them to one particular

◄ Fish supper
The long-winged Arctic tern mostly feeds its young on small fish, crustaceans, and insects, found in shallower water along the shoreline.

Arctic gull
Naturalist John Gould painted this trio of Ross's Gulls, named for British explorer James Clark Ross, for his book *The Birds of Great Britain*, 1862–73.

> "... the white birds were now all flying toward Ahab's boat; and ... began fluttering over the water there, wheeling round and round, with joyous, expectant cries."
>
> HERMAN MELVILLE, *Moby-Dick*, 1851

LESSER BLACK-BACKED GULL
Larus fuscus
Found in the Northern Hemisphere, this large gull has a red spot on its yellow bill. It is often seen scavenging in urban areas.

ROYAL TERN
Thalasseus maximus
This large, orange-billed tern has a black forehead in summer. It is found along North and South American coasts.

AFRICAN SKIMMER
Rynchops flavirostris
A ternlike species, the African Skimmer occurs across sub-Saharan Africa. Adults have an orange bill with a yellow tip.

feeding technique, flying low over water with the elongated tip immersed to detect food, and prevents any kind of foraging.

Originally birds of marshes, watersides, cliffs, and islands, many gulls have now become bold and opportunistic. They exploit various manmade habitats, feeding on garbage dumps or foraging for scraps, even in busy town environments. The flock of gulls following the plows has long been a familiar sight on farmland, and the cheeky gull on the beach has become ever more brazen. In recent decades, species such as the European Herring Gull (*Larus argentatus*) nest increasingly on flat roofs, not only in coastal towns but also far inland. Others, including the Laughing Gull (*Leucophaeus atricilla*) in North America and the Silver Gull (*Chroicocephalus novaehollandiae*) in Australasia, remain firmly associated with their original breeding habitats, such as marshes and islands.

Rare and restricted species

Some gulls are abundant and familiar, whereas others are less common and have particular associations with remote locations. Ross's Gull (*Rhodostethia rosea*), for example, is a rare visitor south of the Arctic, and the Ivory Gull (*Pagophila eburnea*) survives harsh Arctic conditions by feeding on whatever it can find, including seal afterbirths and carcasses of marine mammals and fish. In the Galápagos Islands, the Swallow-tailed Gull (*Creagrus furcatus*) is a tropical species with a particularly appealing elegant form and a sooty-gray hood, set off by a red eye ring. Different leg, bill, and eye ring colors help distinguish gull species, but identification is everywhere confused by the long period—sometimes three or four years—of immaturity, during which the adult plumage develops gradually.

Saving the day

Gulls fly erratically over feeding and roosting sites, flashing white against the sky and visible at long range. This behavior has been exploited by fishermen, who are able to locate surface schools of fish by following the birds. Gulls in general, and the White Tern (*Gygis alba*) in particular, are nicknamed "navigators' friends," because tired or lost sailors are known to follow them back to land. In 1955, the California Gull (*Larus californicus*) was chosen as the state bird of Utah, US. The honor was belatedly bestowed because thousands of gulls ate a plague of crickets that threatened to destroy the local crops in 1848.

Many gull species are migrants, and the Arctic Tern (*Sterna paradisaea*) is often credited with seeing more daylight than any other bird, as it moves between its Arctic breeding grounds and

Antarctic feeding areas. Sea terns, such as the Sandwich Tern (*Thalasseus sandvicensis*), are plunge-divers, hovering above the water before hurtling into the sea to catch fish. The Common Tern (*Sterna hirundo*) plunge-dives into the sea, but over fresh water it tends to dip to the water's surface, like marsh terns. The latter are smaller birds, and they lack the long outer tail streamers that characterize most sea terns. They include the Black Tern (*Chlidonias niger*), which breeds beside fresh water and flies low, head to wind, dipping and swerving to pick up food.

RAPA NUI EGG RACE

In the mythology of the Rapa Nui people of Easter Island, Makemake was a creator god. He was often depicted as a male figure with a bird's head, carrying an egg, and identified as a Sooty Tern (*Onychoprion fuscatus*). The myth tells of an annual race to collect the first egg of the season, the winner of which held power for the coming year.

C.18TH-CENTURY STONE RELIEF

▼ Heading out for food

This dense flock of Black-legged Kittiwakes (*Rissa tridactyla*) is passing a spectacular ice cliff in Svalbard, Norway, as they fly out to sea to catch fish.

Loons and Grebes

Gaviidae, Podicipedidae

Orders Gaviiformes, Podicipediformes

Families 2

Species More than 20

Size range 8–38 in (21–97 cm) long

Distribution Worldwide except Antarctica

These graceful waterbirds, with their beautiful breeding plumages and extravagant courtship displays, are extremely skilled at hunting prey underwater.

Throughout the world, there is a good chance that the local lake or reservoir will be home to a population of grebes. Loons, on the other hand, are solely birds of the Northern Hemisphere. Known as "divers" in Europe, loons are readily associated with the wilderness. The evocative "wailing" call of the Common Loon, or Great Northern Diver (*Gavia immer*), is often dubbed into films and television programs, even those set away from its native range, to create a sense of atmosphere or tension.

Traditionally, loons and grebes were thought to be closely related and were grouped in the same order. However, genetic studies have suggested that this is not the case at all. In fact, their relationship seemingly goes no deeper than sharing similar physical attributes, such as a streamlined body and large webbed feet for diving for fish, aquatic insects, and sometimes amphibians. Regardless, both families are skilled nest builders, with grebes constructing floating nests from dead and rotting plant matter, and

◄ **Bow to your partner** Great Crested Grebes (*Podiceps cristatus*) swim toward each other slowly, in silence, as part of their long courtship ritual. They then continue the display with a series of dance moves, shallow dives, and mating calls.

loons using the same material to create large platforms in shallow water. In winter, many species swap freshwater for marine habitats, and so can be found off coasts and in bays.

Both loons and grebes change their plumages dramatically between seasons. In winter, most species show subdued tones of black, gray, brown, or white, but in summer they look far more striking. Grebes are renowned for their elaborate head plumages and bright coloring.

Legendary loons

Loons have long been prominent in North American Indigenous culture. They feature in creation stories, including those of the Sioux, in which the loon was one of the animals enlisted to dive to the ocean floor and bring up earth to create land. More contemporarily, the Canadian one dollar coin has affectionately become known as the "loonie" because it features a common loon on one side.

▲ **Strong swimmers**
As this 1833 watercolor by Audubon shows, the legs of a Red-throated Loon (*G. stellata*) are positioned far back on the body, making them excellent swimmers but ungainly out of water.

BLACK-THROATED LOON
Gavia arctica
Able to dive to depths of 20 ft (6 m), this species breeds in Arctic and sub-Arctic regions across Eurasia, wintering in sheltered coastal areas.

" Even the shores seemed hushed and waiting for the first lone call . . . "

SIGURD F. OLSON, *The Singing Wilderness*, 1956

Water taxi
A common loon keeps its chick warm and safe from predators by giving it a "piggyback" ride across the water.

LITTLE GREBE
Tachybaptus rufolavatus
Familiar throughout Europe, Africa, and Asia, the seven subspecies of the Little Grebe differ in size, plumage, and eye color.

TITICACA GREBE
Rollandia microptera
This short-winged flightless species is found mainly at Lake Titicaca on the border of Peru and Bolivia. It is a superb and speedy diver.

◄ Speed waddlers
In early spring, Adélie penguins may walk 20 miles (30 km) or more across an ice sheet to reach their nests, averaging speeds of 1½ mph (2.5 kph).

Strong claws grip the ice

Outstretched flippers help with balance when walking on slippery ice

Order Sphenisciformes

Families 1

Species More than 10

Size range 1⅓–3¾ ft (40–115 cm) tall

Distribution Antarctica, Southwestern Africa, South America, Australasia, Southern Pacific islands

> " One can't be angry when one looks at a penguin. "
>
> JOHN RUSKIN, Letter to US scholar Charles Eliot Norton, 1860

Penguins

Spheniscidae

Famously flightless, penguins are found in cold marine habitats. Their adaptation to life at sea surpasses all other birds, with their wings acting as flippers to power them at speed through the water.

Back feathers range from slate grey to blue, leading to little penguins sometimes being called blue penguins

▲ South Seas study
This Little Penguin (*Eudyptula minor*) was sketched by George Forster in 1773. As its name suggests, it is the smallest species of penguin.

Adored for their characterful waddle and fluffy chicks, penguins are highly social and many species gather in large breeding colonies. They live almost exclusively in the Southern Hemisphere and are adapted to extremely cold habitats—temperatures drop to -76°F (-60°C) during Antarctic winters. Only the Galápagos Penguin (*Spheniscus mendiculus*) breeds on the equator, where the cold Humboldt Current brings nutrient-rich water from the Southern Ocean to the islands.

Penguins spend much of their lives at sea, only coming ashore to breed or molt. To survive in the cold, penguins have a thick, insulating layer of body fat, and their feathers are densely and uniformly packed across the body, which provides more insulation and waterproofing. The feathers are replaced annually by a "catastrophic

◄ Life on ice
After eight to nine weeks of incubating an egg on his feet, the male emperor penguin is joined by the female, who provides food for their hungry chick.

molt," where new feathers grow and push out the old ones all at once. The whole process takes around 3–4 weeks and during this time the penguin cannot go to sea to feed until its new waterproof coat is ready.

Immense colonies

Emperor Penguins (*Aptenodytes forsteri*) breed in large colonies on the frozen Antarctic ice. Each pair lays a single egg, which the male incubates alone throughout the harsh winter. Most species of penguins, however, lay two eggs. For crested penguins (*Eudyptes* spp.) only the much larger second egg hatches and scientists think this may be linked to the first egg developing while the female is still at sea.

Adélie Penguins (*Pygoscelis adeliae*) form colonies of as many as a million breeding pairs. Like Gentoo (*P. papua*) and Chinstrap (*P. antarcticus*) penguins, they breed on ice-free land on Antarctica where they make raised pebble nests, which keep the eggs above any melting snow. They are renowned for stealing pebbles from other nests or even settling into another penguin's abandoned nest.

King Penguins (*A. patagonicus*) raise each chick for over a year, often rearing two chicks in a three-year period. Because of this, their colonies

WARM BODY, COLD FEET

Penguins have a special countercurrent circulatory system, which conserves body heat by reducing heat loss from cold extremities. Warm blood flowing to a foot transfers heat to the cold blood returning to the body as it passes close by. This keeps the feet cool, and the penguin's core warm.

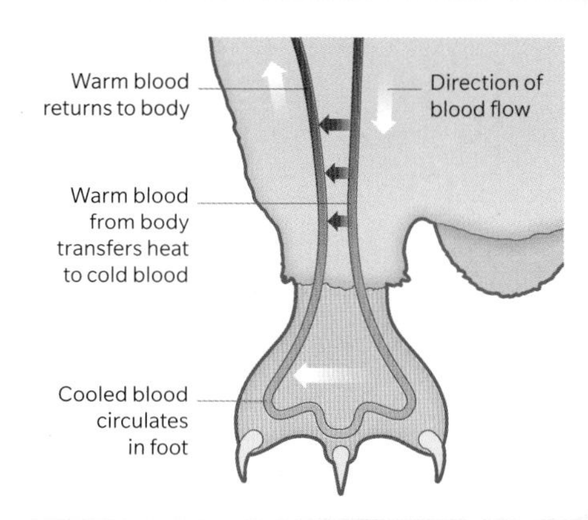

COUNTERCURRENT SYSTEM

"... there are birds as big as ducks, but they cannot fly ... and they bray like donkeys."

VASCO DA GAMA, *A Journal of the First Voyage of Vasco da Gama*, 1497

◄ **Forest breeders**
These Fiordland Penguins (*Eudyptes pachyrhynchus*) are returning from the sea to their nests, which are burrows dug into the coastal forest floor of New Zealand's South Island.

▲ Plunging in
Gentoo penguins jump from ice floes, often spending the entire day hunting for krill. They are the fastest bird underwater, swimming at speeds of up to 22 mph (36 kph).

usually contain lots of fluffy brown chicks of different ages. These early brown feathers, which are not waterproof, must molt and be replaced before the chick can go into the sea to hunt for food itself.

Many chicks gather in crèches when they are about 3–4 weeks old. Crèches enable both parents to hunt for food, for themselves and their chick, while the chick remains safe in a huddle with other chicks and guarded by a few adults. However, this can only occur once the chick can regulate its body temperature, and parents are able recognize their chick by its calls.

Consummate swimmers

Penguins have a streamlined body and webbed feet that help them swim through water quickly. They also have the densest bones of any bird, enabling them to swim deep underwater in search of fish, krill, and squid. In one dive, an emperor penguin reached the incredible depth of 1,850 ft (565 m). Penguin wing bones are flattened and the joints are fused together to form a stiff paddle that cannot be held against the body. This helps penguins balance on land and provides power underwater.

Many species of penguin porpoise, leaping out of the water to breathe before diving back below the surface. Porpoising allows them to swim faster and often occurs near a colony, where the threat from predators, such as leopard seals, is highest.

PENGUIN BOOKS
THE BODLEY HEAD
GONE TO EARTH
MARY WEBB
THE BODLEY HEAD
COMPLETE
UNABRIDGED

Striking orange bands on the cover indicated that the paperback was a work of fiction

◄ Black-and-white icon
The appeal of penguins has seen them feature culturally in many ways, including as the logo of Penguin Books. The first logo was designed in 1935 and has since been redesigned several times.

SOUTHERN ROCKHOPPER PENGUIN
Eudyptes chrysocome
One of the crested penguins, this species mainly nests high on rocky slopes. Many have to leap from heavy surf onto a rocky shore and then climb by jumping.

HUMBOLDT PENGUIN
Spheniscus humboldti
Found along the west coast of South America, this species breeds in desert habitats adjacent to the cold Humboldt Current. Patches of bare skin help it stay cool in summer.

AFRICAN PENGUIN
Spheniscus demersus
Also known as the jackass penguin due to its braying call, this endangered species has colonies on the coasts of southwestern Africa, around South Africa and Namibia.

Diving for dinner
Having dived into the cold waters of the Southern Ocean, these gentoo penguins are racing to hunt for prey, such as krill, small fish, octopus, or squid. Gentoos are the fastest underwater swimmers of all the penguins, flapping their stiff wings through a wide arc and steering with their feet.

Albatrosses

Diomedeidae

Emblematic of freedom as they soar over the vast, windswept oceans of the Southern Hemisphere for months at a time, albatrosses are famously long-winged and include some of the world's largest flying birds.

Order Procellariiformes

Families 1

Species More than 20

Size range 2½–4⅓ft (75cm–1.35m) long

Distribution South America, Africa, Australasia, Antarctica, Southern Ocean

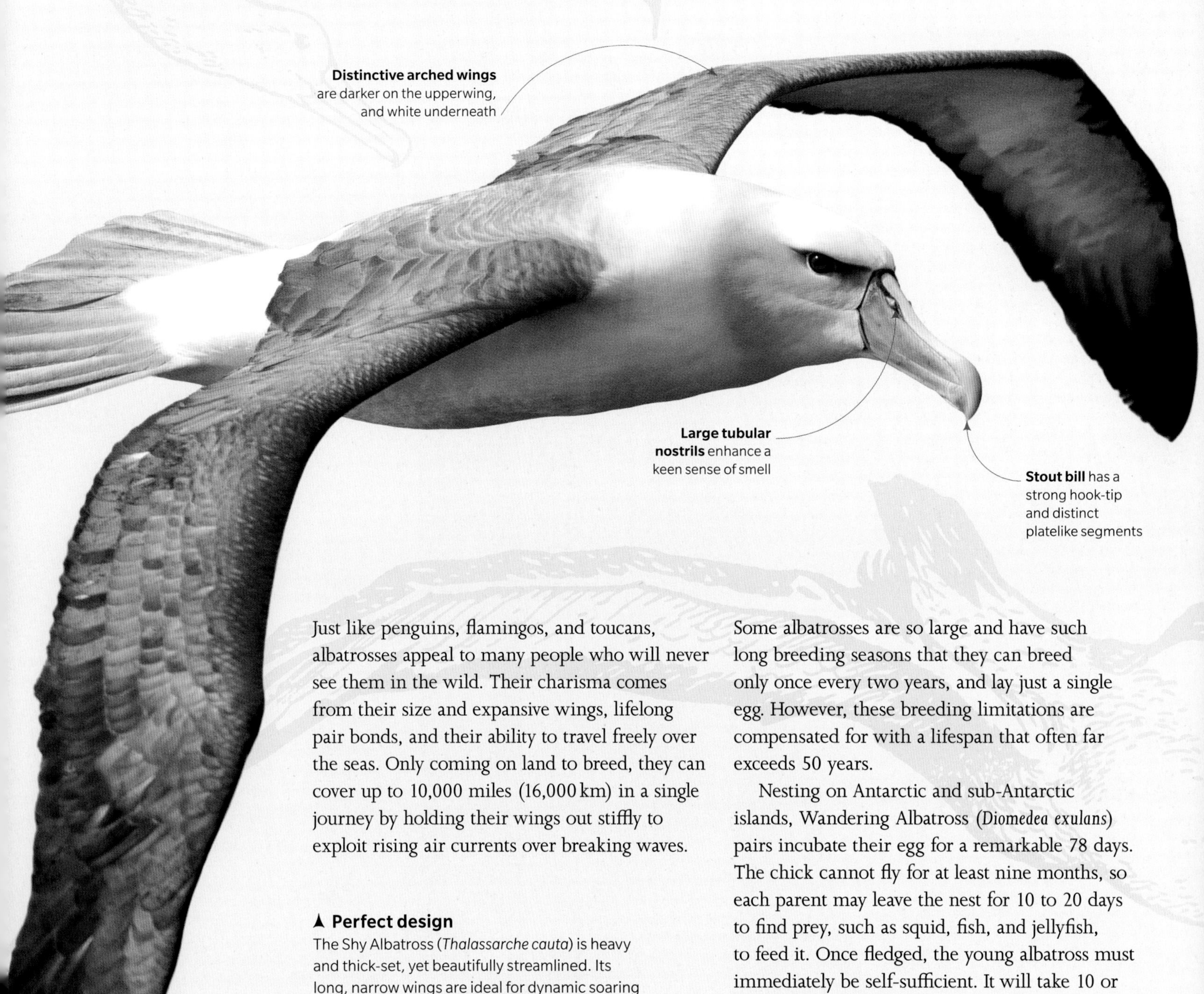

Just like penguins, flamingos, and toucans, albatrosses appeal to many people who will never see them in the wild. Their charisma comes from their size and expansive wings, lifelong pair bonds, and their ability to travel freely over the seas. Only coming on land to breed, they can cover up to 10,000 miles (16,000 km) in a single journey by holding their wings out stiffly to exploit rising air currents over breaking waves.

▲ Perfect design
The Shy Albatross (*Thalassarche cauta*) is heavy and thick-set, yet beautifully streamlined. Its long, narrow wings are ideal for dynamic soaring on complex air currents over wave crests.

Some albatrosses are so large and have such long breeding seasons that they can breed only once every two years, and lay just a single egg. However, these breeding limitations are compensated for with a lifespan that often far exceeds 50 years.

Nesting on Antarctic and sub-Antarctic islands, Wandering Albatross (*Diomedea exulans*) pairs incubate their egg for a remarkable 78 days. The chick cannot fly for at least nine months, so each parent may leave the nest for 10 to 20 days to find prey, such as squid, fish, and jellyfish, to feed it. Once fledged, the young albatross must immediately be self-sufficient. It will take 10 or more years before it is ready to breed.

Albatrosses have a remarkable sense of smell and excellent navigational skills, returning to their small island colony with ease, and sailors long believed that albatrosses would help them navigate away from hazardous weather conditions. They became a symbol of good luck and simply spotting one at sea was considered an indicator of a safe voyage ahead. It was even thought that albatrosses embodied the souls of dead sailors and therefore should not be harmed. Only with the poem *The Rime of the Ancient Mariner* by Samuel Taylor Coleridge in the late 18th century did the bird become associated with regret and misfortune.

Cursed ship
The Rime of the Ancient Mariner tells of a ship that is cursed when a sailor kills an albatross. The dead bird around his neck is a motif for regret.

> " Instead of the cross, the Albatross about my neck was hung. "
>
> SAMUEL TAYLOR COLERIDGE,
> *The Rime of the Ancient Mariner, 1798*

Albatrosses have also been associated with power due to their incredible wingspan. The family name Diomedeidae is derived from Diomedes, a hero and commander of 80 ships in Greek mythology. In one story, his companions are transformed into birds on a voyage from Argos.

Exploring the north

While most albatrosses remain in the Southern Hemisphere, a few penetrate calm-air equatorial regions and some have even visited Northern Gannet (*Morus bassanus*) colonies in the North Atlantic and North Sea, to the delight of many birdwatchers. They have usually been blown off course by heavy storms, and follow oceanic trade winds into the Northern Hemisphere.

Many albatross colonies have lost large numbers of birds because they swallowed baits on hooks from longline fishing vessels. However, the populations of some species, including the Black-browed Albatross (*Thalassarche melanophris*), are increasing due to conservation initiatives, such as protected breeding sites and measures to reduce the number of seabirds killed by fishing.

▲ Nosy neighbors
The black-browed albatross builds its basinlike nest on coastal cliffs, among other bird colonies. Although the neighboring penguins pose little threat, the adult albatross warns them away from its egg.

WANDERING ALBATROSS
Diomedea exulans
With a wingspan of up to 12 ft (3.6 m)—the largest of any living bird—this species travels thousands of miles with little energy expenditure.

WAVED ALBATROSS
Phoebastria irrorata
This unusual tropical albatross nests on the equator in the Galápagos Islands. It feeds on fish and squid, which are often caught at night.

Petrels and Shearwaters

Oceanitidae, Hydrobatidae, Procellariidae

Order Procellariiformes

Families 3

Species More than 100

Size range 6–37 in (16–93 cm)

Distribution Worldwide

Like their albatross relatives, these seabirds are ocean voyagers, guided as much by their nostrils as their eyes. On land, however, they are vulnerable to attack from rats and other predators.

Petrels and shearwaters roam huge distances across the world's oceans when not breeding. Their keen sense of smell, enhanced by tube-shaped nostrils, enables them to find floating carrion and other food, and also track their way back to the remote coasts and islands where they nest. Shearwaters and gadfly petrels have a powerful scything flight on long, slender wings, while the tiny, broader-winged storm petrels flutter, dangling their long legs and pattering their webbed toes on the water (hence the name "petrel"—after St. Peter, the disciple of Jesus who walked on water). The diving petrels look and behave more like auks, diving for food rather than picking it from the surface.

Spirits of the storm

The name Procellariiformes is derived from *procella*—Latin for a furious storm—and even the smallest of these birds can handle heavy weather at sea. Accordingly, English seafarers nickname the tiny European Storm Petrel (*Hydrobates pelagicus*) "Mother Carey's chicken," Mother Carey being a ship-sinking sea-witch in 18th- and 19th-century sailors' folk tales. The burly, rather gull-like giant petrels of the Antarctic are sometimes known as "Mother Carey's geese."

Shearwaters and storm petrels fly and swim very well, but are awkward on land as they cannot raise their ankles off the ground. This makes them vulnerable to predators such as gulls. Most smaller species nest in burrows and visit their colonies at night, to avoid such hunters. Against predatory mammals, they are almost defenseless, and the introduction of species such as rats to their islands is disastrous. Fifteen species are critically endangered, and efforts to save them focus on eradicating nonnative predators, or relocating birds to predator-free islands. However, with luck, even small storm petrels can live for more than 30 years.

Storm raiser
This illustration from 1877 shows storm petrels as the souls of sailors flying alongside the storm-raising sea-witch Mother Carey.

The Palawa people have hunted shearwaters for meat and oil in Tasmania for at least 8,000 years

◄ Running start
The Sooty Shearwater (*Ardenna grisea*) has long, narrow wings ideal for gliding on its 40,000-mile (64,000-km) migratory journey. But it needs a running start to get airborne from the water.

◄ Fulmar frenzy
Northern Fulmars (*Fulmarus glacialis*) fly solo for much of the time, but large numbers converge on a particularly bountiful food source, such as a floating whale carcass.

SOUTHERN GIANT PETREL
Macronectes giganteus

At up to 3.3 ft (1 m) long, this Antarctic petrel has little to fear from predators and is itself a powerful and aggressive hunter of other seabirds.

SNOW PETREL
Pagodroma nivea

This pure white seabird will nest far inland on Antarctica. It is one of only three bird species recorded at the geographic South Pole.

WILSON'S STORM PETREL
Oceanites oceanicus

This diminutive storm petrel breeds across sub-Antarctic islands and coastal Antarctica, but wanders much farther north outside the breeding season.

The bright yellow frontal shield, or saddle, and the bare red skin may help the stork keep cool

▲ Soaring high
The saddle-billed stork can reach more than 5 ft (1.5 m) in height in captivity, making it one of the tallest stork species. It lives throughout the tropics in Africa.

Storks

Ciconiidae

With their long legs, prominent powerful bill, and bold plumage, storks are easy to identify. Long associated with good fortune, these large birds are the source of many myths.

Order Ciconiiformes

Families 1

Species 20

Size range 27–60 in (68–152 cm) long

Distribution Worldwide except Antarctica

Storks are large, long-legged birds that live in freshwater habitats, such as swamps and marshes, or open areas, such as grasslands. They are among the most of striking birds, with their long, heavy bills—the Marabou Stork (*Leptoptilos crumenifer*) has the largest, at nearly 14 in (35 cm). Most have stark, black-and-white plumage, although some have some gray or pinkish feathers too. The African Openbill (*Anastomus lamelligerus*) is unusual, having long, ribbonlike feathers with a greenish sheen on its front and back. In most storks, the males and females are identical, but the female Saddle-billed Stork (*Ephippiorhynchus senegalensis*) has a yellow iris, while the male has a brown one.

Many species have patches of bare skin on their head that become more vividly colored during the breeding season. For example, Abdim's Stork (*Ciconia abdimii*) develops red skin in front of its eyes, and blue skin on its face when it is ready to breed. The bare skin on the heads of the marabou stork and Greater Adjutant (*L. dubius*), meanwhile, serves another purpose: it helps the birds keep clean when feeding on carcasses.

Migration mystery solved

One of the most remarkable discoveries about bird migration was made by observing a wounded stork. In 1822, a White Stork (*C. ciconia*)

Long, broad wings enable the stork to take off quickly with sweeping wingflaps and soar effortlessly on thermals

➤ Woodland mystery
This fanciful art nouveau illustration by Otto Eckmann for a 1898 German magazine cover depicts a group of puzzled Painted Storks (*Mycteria leucocephala*) gathered around an egg in a forest.

Several species of stork defecate onto their legs, which keeps them cool by evaporation

was seen in a German village with a 30 in (75 cm) spear impaled in its neck. Until this time, naturalists had been uncertain as to where some species of storks disappeared to in the winter—they had even wondered if certain species hibernated under the ground. The spear was made of a dark wood of African origin and was the first proof that birds migrated between Europe and Africa. In the following years, at least 25 of these *pfeilstörche* or "arrow storks" would be seen in Germany.

White storks follow two migration routes into Africa, one crossing the Mediterranean at Gibraltar and the other passing across the Bosphorus in Turkey. All storks fly with their necks outstretched, like cranes, apart from marabou and the two adjutant storks, which fly with their necks retracted, like herons. The long broad wings of storks allow them to soar well on thermals, and so travel distances with ease. The sight of thousands of migrating white storks soaring high is wonderful to behold.

Marabou storks and greater adjutants have inflatable air sacs that hang from their throats. They also have a second, smaller air sac at the base of the hind neck. The inflation of the throat sac displays dominance, while an inflated neck sac

Naughty babies are carried in the bill; good babies are brought in a cloth bundle or basket

▲ Delivering a baby
The European myth of white storks delivering babies has its origins in ancient Greece, but it was made popular by a Hans Christian Andersen fairy tale written in 1839.

▲ A welcome return
In Europe, white storks are encouraged to nest on buildings to bring luck. Here, a chick looks on as its parents clatter their bills and throw back their heads in welcome as one returns to the nest.

denotes apprehension. Both may be inflated during courtship. The throat sac is connected to the left nostril and produces a guttural croaking during display. It also helps to cool the birds.

No place like home

Most storks nest in colonies in trees, often alonside other stork species and sometimes with herons, egrets, ibises, or cormorants. Maguari Storks (*C. maguari*) nest on the ground in reed beds, while Abdim's stork and the Lesser Adjutant (*L. javanicus*) nest in cliffs and on rock pinnacles. The saddle-billed storks and jabiru are solitary nesters, and they also tend to stay together as a pair after breeding, unlike most other storks. Nests are usually built of sticks, and the white stork reuses its nests, adding more sticks each year, with some becoming at least 10 ft (3 m) deep. Storks are not very vocal, as they do not have a fully developed syrinx, or voice box. They can produce croaks, hisses, and whistles but mainly communicate by clattering their bills. Most storks do this by hitting the upper and lower mandibles together, but openbills do it differently—the bird clatters its bill against that of its partner.

In ancient Greece, young storks were believed to look after elderly parents, and so in heraldry, the stork represents filial duty and gratitude.

> " In Thessaly it is regarded as a capital crime to kill a stork. "
>
> PLINY THE ELDER, *Naturalis Historia*, 77 CE

AFRICAN OPENBILL
Anastomus lamelligerus
Young openbills have a straight bill like other storks. The characteristic curved opening takes several years to fully develop.

GREATER ADJUTANT
Leptoptilos dubius
Once endangered, this large stork from northern India and Cambodia has seen its population grow thanks to conservationists.

JABIRU
Jabiru mycteria
The Jabiru's wingspan, measuring about 9 ft (2.8 m), is the second largest in South America, after that of the Andean Condor (*Vultur gryphus*).

Traditional bird hunters of Bihar, India, practiced a ritual that required a young man to capture a Black-necked Stork (*E. asiaticus*) alive with a stick covered in glue before he could marry. The ritual was stopped in the 1920s after a man was killed by the bird. In West Africa, various parts of the marabou stork, such as the air sacs and intestines, were used in remedies for rheumatism, memory loss, and other ailments. The bird's skin was even used as a vest to protect against witchcraft.

Snappy eaters

Storks have a wide diet, feeding on insects, fish, amphibians, reptiles, and small mammals. The marabou stork and greater adjutant are largely carrion feeders, Abdim's stork eats mainly insects, while openbills love large aquatic snails.

Storks of the *Mycteria* genus have a down-curved bill with a sensitive tip that enables them to feed without seeing their prey—which they capture using a technique known as grope-feeding. Feeding often takes place in water 6–10 in (15–25 cm) deep, where a stork will probe with its bill partly open. When a fish touches the bill, it quickly snaps shut. The average response time of this reflex in the Wood Stork (*M. americana*) is 25 milliseconds, making it one of the fastest reflexes known in vertebrates.

The **average weight** of a white stork's **nest** is about **900 lb** (400 kg), with some weighing over **1 ton**

➤ Lakeside colony
The Sultanpur National Park wetlands in northern India are home to resident colonies of painted storks, attracted by the abundant fish and frogs, and suitable trees for nesting.

Frigatebirds

Fregatidae

Extremely slender, angular wings tapered to a long point and a deeply forked tail give the frigatebirds an almost prehistoric look.

Inlays of mother of pearl and nautilus shell form an intricate pattern

▲ **Ceremonial bowl**
This early 20th century wooden bowl carved in the form of a frigatebird may have been used in initiation rites and feasts in the Solomon Islands in the South Pacific.

Remarkably for birds so strictly associated with the sea, frigatebirds cannot settle on the water surface. Their plumage is not waterproof, and their feet and legs are so feeble that they cannot kick free of the water nor run along the surface to gain speed for takeoff. Instead, frigatebirds rely on supreme flying abilities to exploit wind currents and travel vast distances over the ocean, using very little energy, without having to rest on the water at all. They feed in flight, taking fish and squid from near the surface or stealing food from other seabirds.

Charismatic gliders

Their wings are particularly long and slender, the ideal shape for gliding over great distances, and the forked tail can be spread wide or closed to a point to assist tight maneuvers. The wings also by chance give a superbly charismatic and impressive appearance, which has long caught the attention of mariners and explorers. The common name frigatebird has been used since at least the 18th century, although many early sailors called them man o' war birds or, in Old Spanish, *rabiforçado* (forktails).

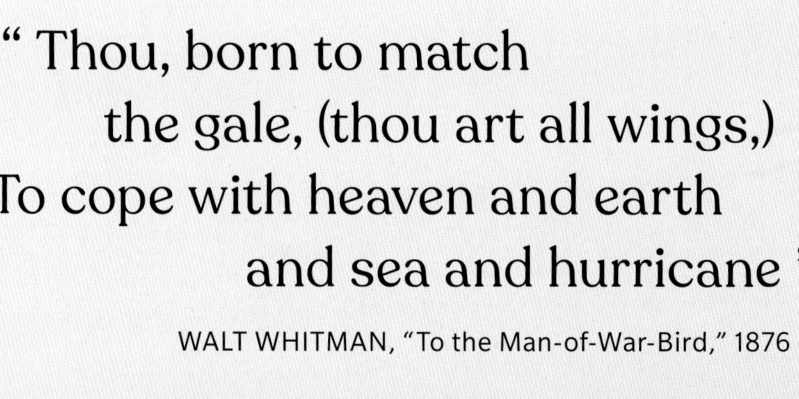

> " Thou, born to match the gale, (thou art all wings,) To cope with heaven and earth and sea and hurricane "

WALT WHITMAN, "To the Man-of-War-Bird," 1876

➤ **Mugging a pelican**
It takes a matter of moments for a Brown Pelican (*Pelecanus occidentalis*) to swallow a fish, but an immature Magnificent Frigatebird, despite its size, is so agile it can snatch the prize from inside the pelican's pouch.

A **juvenile frigatebird** may **wander** as far as **3,700 miles** (6,000 km) from the island **where it was hatched**

Colonies usually form on islands with bushy trees or rocky outcrops. Males inflate a red throat pouch and sit on open perches in a thicket, showing their finery to other males and females flying over in a display of fitness to breed and father a healthy chick. Females lay just a single egg every two years, as the egg is incubated for more than 40 days and the chick may be fed at the nest for up to six months in the larger species. It may be a further 10 years before the young bird can breed. A colony may easily be disrupted by disturbance, predators, and over-fishing, and the low breeding rate means it can be difficult for a population to recover.

Nevertheless, frigatebirds remain easy to see, especially in the Caribbean Sea, where they soar over harbors in cities such as Havana, Cuba, or around island groups such as the Seychelles in the Indian Ocean. They are also often spotted diving for swarms of small fish being hauled onto a beach, almost within reach of the fishermen and holidaymakers.

Order Suliformes

Families 1

Species 5

Size range 29½–45 in (75–115 cm) long

Distribution North America, South America, Africa, Asia, Australasia

➤ Bright red balloon

All male frigatebirds, such as this Magnificent Frigatebird (*Fregata magnificens*), have a throat pouch—called a gular sac—which they inflate for courtship and territorial displays.

The adult has an all-black head, having had a largely white neck and head as a juvenile

The slender, hook-tipped bill is ideal for snatching prey, such as flying fish, from the wavetops, or fish dropped by other birds

The fleshy throat sac under the throat turns into an extraordinary "balloon" when inflated

Boobies and Gannets

Sulidae

▲ **High-speed diving**
Where fish are abundant, dozens of northern gannets may plunge-dive in close proximity to one another, a risky business that can occasionally result in injury.

Among the largest of the seabirds, boobies and gannets perform impressive headlong plunge-dives in pursuit of fish. They build a lifelong bond with their partner, affirmed through displays and dances.

Order Suliformes

Families 1

Species 10

Size range 27½–37½ in (70–95 cm) long

Distribution Worldwide

When a booby or gannet folds in its wings and punctures the sea like a feathered javelin, it is every inch the efficient predator. Physical adaptations that allow face-first diving at speeds of 60 mph (95 kph) include a highly streamlined body shape, nostrils that open inside the bill, and sturdy nictitating membranes (third eyelids) to protect the eyes. Totipalmate feet (with all four toes connected by webbing) enable the birds to swim well underwater and on the surface. Having seized a fish in their bill, they rapidly swallow it whole. This is not done through greed, but because other birds might attempt to steal their catch. On land, boobies and gannets had little to fear until humans first made landfall on the birds' previously predator-free breeding islands, such as the Galápagos in the 1500s. Historically, gannet and booby chicks were an important traditional food source in many Pacific and North Atlantic island communities, and some hunting is still practiced today. The Red-footed Booby (*Sula sula*) is still hunted in the Cocos Islands, for example, and in the UK, a legal hunt of Northern Gannets (*Morus bassanus*) is held annually on the uninhabited and perilously craggy Scottish island of Sula Sgeir, although the hunters are only permitted to take a few fledglings or "gugas."

GUANO MINING

The Peruvian booby (*Sula variegata*) is a globally important "guano bird," and deposits of its droppings have long been mined for valuable fertilizer. The "guano age" (1802–84) saw heavy exploitation of guano on islands on the Pacific coast of South America. The Cape Gannet (*Morus capensis*) is another important source of guano. This species has declined steeply in Namibia since the 1950s, but it is increasing in South Africa.

CHINCHA ISLANDS, PERU, 1865

Lifelong partners

All boobies and gannets are long-distance wanderers and may travel thousands of miles from their birth colony when not breeding. They can live for more than three decades, and will reunite with the same partner every year after spending the nonbreeding season at sea alone. In courtship, pairs display through clattering their bills together, or in some species showing off their brightly colored feet—a reliable indicator of physical health. Most booby species produce two eggs per breeding attempt, but often only one chick survives, because the older one kills its younger sibling. When it fledges, the chick is heavier than its parents, and that extra weight helps sustain it when it must take flight and learn to feed itself.

AUSTRALASIAN GANNET
Morus serrator

This gannet breeds in Australia and New Zealand. Conservationists use decoys to encourage wandering birds to establish colonies on new islands.

MASKED BOOBY
Sula dactylatra

The world's largest booby breeds in dense colonies on remote tropical and subtropical oceanic islands and coral atolls.

> " [The booby] is a very simple creature, and will hardly go out of a man's way. "
>
> WILLIAM DAMPIER, *A New Voyage Round the World*, 1697

◄ Showing off
The Blue-footed Booby (*Sula nebouxii*) breeds on the Galápagos Islands, where visitors enjoy its pair-bonding display.

Cormorants and Darters

Phalacrocoracidae, Anhingidae

Famed for their prehistoric looks and adept fishing techniques, these are characteristic birds of aquatic environments. They have a long and varied relationship with humans.

◀ Drying out
Cormorant feathers absorb water, helping the bird sink when diving for fish. This great cormorant has returned to shore to dry its plumage.

Order Suliformes

Families 2

Species More than 40

Size range 18–39 in (45–100 cm) long

Distribution Worldwide

Although many cormorant and shag species may seem slightly clumsy when observed out of water, they are highly skilled at seeking fish beneath the surface. Using their webbed feet to propel them, cormorants can stay underwater for two minutes or more and some species have been recorded at depths in excess of 200 ft (60 m).

Their close cousins, the anhingas and darters, are often referred to as "snakebirds" due to their long, thin neck that often sticks out of the water while the rest of the body is submerged, giving them a serpentine appearance. They use their daggerlike bill to harpoon their prey.

▲ Ancient fishing tradition
A 19th-century Japanese woodblock print by the artist Keisai Eisen shows fishermen employing cormorants to catch fish on the Nagara River. This fishing technique is still used in China, Japan, and Peru.

Mixed reception

The underwater prowess of cormorants has won them both friends and foes among humans. Although close working relationships have been formed with the birds in some cultures, in other quarters they are regarded as unwelcome competition because of a perceived threat to fish stocks—cormorants have been seen fishing in flocks thousands strong. Populations of some species, such as the Great Cormorant (*Phalacrocorax carbo*) in Europe, were historically persecuted to the point of extinction, although they have since recovered. Cormorants feature both positively and negatively in symbolism worldwide, having been viewed as icons of greed, gluttony, bad luck, and evil, but alternatively also being recognized as a positive omen of good luck, dedication, and reliability.

Economic value

Cormorant guano is an excellent fertilizer due to its high content of potassium, nitrogen, and phosphate. An industry once existed around the exportation worldwide of guano from the Guanay Cormorant (*Leucocarbo bougainvillii*), native to the coasts of Peru and Chile. This industry peaked during the 19th century but rapidly declined in the early 20th century as the Haber-Bosch process was invented, allowing nitrogen to be extracted from the atmosphere to produce synthetic ammonia.

Cormorant comes from the Latin *corvus marinus*, which means "sea crow"

AFRICAN DARTER
Anhinga rufa
Male African Darters have a two-toned rufous throat and neck with elongated white plumes forming a facial stripe in breeding season. The species is widespread across sub-Saharan Africa.

FLIGHTLESS CORMORANT
Nannopterum harrisi
Restricted to Fernandina and Isabela Islands in the Galápagos, this is the only flightless species in the family. Its wings are used for maneuvering in water while fishing.

Team work
As dusk falls, a fisherman takes his great cormorants out on the Li River at Yangshuo, Guangxi, China, continuing a tradition that began more than 1,000 years ago. He will tie a short piece of string loosely around each bird's neck so that any large fish they catch will be held in their gullet but leaving them free to swallow smaller fish.

Pelicans

Pelecanidae

Imposing figures at rivers, lakes, and coastal waters, pelicans are prolific hunters of fish, using an elastic pouch under their remarkably long bill to net their prey in great quantities.

Order Pelecaniformes

Families 1

Species 8

Size range 4–6¼ ft (1.2–1.9 m) long

Distribution Worldwide except Antarctica

Short-legged and clumsy on land, pelicans are buoyant on water and a magnificent sight in the air, despite being so heavy. They use their huge wings to soar on thermals to a great height before gliding away to distant feeding sites, or traveling on their long migrations. European Great White Pelicans (*Pelecanus onocrotalus*) migrate to East Africa, where colonies number in the tens of thousands. Each bird needs to consume around 2 lb (1 kg) or more of fish a day, meaning they can survive in only the richest wetlands.

While many birds have a flexible, fleshy sac between the bill and the throat, this gular pouch is especially developed in pelicans, holding up to 3½ liquid gallons (13 liters) of water. The large, sensitive bill detects fish by touch. They are scooped up and the lower mandible forms a hoop above the wide-open pouch, from which water spills out sideways before the fish are swallowed whole.

Blood sacrifice

Ancient legends claiming that in times of famine pelicans fed their young with their own blood—or even that the parent brought their chick back from the dead with their blood, sacrificing themselves in the process—were readily adopted by early Christians. The myth was likened to the sufferings of Jesus and came to symbolize Christian charity.

◄ Immense hunter
Arguably the world's largest wetland bird, the Dalmatian Pelican (*Pelecanus crispus*) has a wingspan of 11½ ft (3.5 m) and a bill up to 18 in (45 cm) long. To fuel its big body, it drives shoals of fish into the shallows then swallows them in great gulps.

Christian symbol
This mosaic in the Kykkos monastery, Cyprus, depicts the sacrifice of a mother pelican piercing her breast to feed her young.

> "... thus wide
> I'll open my arms,
> And, like the kind
> life-rend'ring pelican,
> Repast them with
> my blood."
>
> WILLIAM SHAKESPEARE, *Hamlet*, c.1600

AUSTRALIAN PELICAN
Pelecanus conspicillatus
A familiar sight on Australian beaches and lakes, this often remarkably tame pelican breeds in colonies of thousands whenever conditions become suitable following rainfall.

BROWN PELICAN
Pelecanus occidentalis
Unusual among pelicans, this medium-sized American species plunge-dives headlong into coastal waters, stunning fish with the impact before scooping them up in its pouch.

PINK-BACKED PELICAN
Pelecanus rufescens
This smaller species still has a large wingspan of up to 9½ ft (2.9 m). It is resident in Sub-Saharan Africa and the southwestern Arabian Peninsula at freshwater lakes, swamps, and rivers.

"... the most extraordinary bird
I have seen for many years ..."

JOHN GOULD, *Proceedings of the Zoological Society of London*, 1851

➤ **Slow flaps**
It may appear top-heavy, but the shoebill is a good flier. Like herons, it flies with its neck retracted, and its large wings mean it can soar well on thermals.

Broad wings flap slowly when the shoebill flies over water and reedbeds

Long legs trail behind during flight

Shoebill

Balaenicipitidae

The shoebill stands alone in its own family, quite unlike any other bird with its outsized, hook-tipped, bulbous bill and large, staring pale eyes.

Unmistakeable with its huge clog-shaped beak, long legs, and blue-gray plumage, the Shoebill (*Balaeniceps rex*) is a large bird that inhabits African freshwater swamps. Its long toes help it to walk over submerged or floating vegetation despite its large size.

Taxonomic oddity

The shoebill was depicted by ancient Egyptians and petroglyphs (prehistoric rock art) also exist in Algeria. Once descriptively named the whale-headed stork, its taxonomic position has long been a puzzle but recent DNA studies have shown it to be most closely related to pelicans and the equally unusual hamerkop.

Shoebills are monogamous, and a breeding pair builds a large nest surrounded by water and at times even on floating vegetation. Of the one to three eggs laid, only one chick usually survives. In very hot weather, parents fill their beak with water and then use this to cool the eggs or chicks.

Favorite foods include lungfish and catfish, but the shoebill will also eat waterbirds, water snakes, amphibians, and even small crocodiles. It is a stealthy hunter, moving slowly and standing still for long periods of time, resting its bill on its chest, watching for movements beneath the water. On locating prey, it will lunge forward and grab a huge mouthful of water, mud, vegetation, and hopefully its meal.

SHOE-BILL

▲ **Trading card**
This US cigarette trading card from 1889 was part of a series of 50 cards called "Birds of the Tropics."

Order Pelecaniformes

Families 1

Species 1

Size range 4 ft (1.2 m) tall

Distribution Northern, Central, Southern Africa

Hamerkop

Scopidae

This bizarre-looking evolutionary puzzle builds the largest nest of any bird, which it decorates with strange objects, prompting an association with magic.

There is, literally, no bird like the Hamerkop (*Scopus umbretta*). The only species in its family, the hamerkop is so evolutionarily unique that scientists have struggled to determine its nearest relative. Although hamerkops recall miniature storks, current thinking plumps for the equally unusual shoebill and the pelicans.

The hamerkop's enormous nest contains a single entrance tunnel on the underside, a thick roof, and internal walls that are plastered with mud. Remarkably, a pair may build a handful of nests each year and use up to a dozen on rotation. Pairs also decorate their nests with ornaments ranging from animal feces, bones, and other birds' feathers to human clothing and household objects.

This practice recalls that of African shamans, and so this particular idiosyncratic behavior has inspired the belief that the hamerkop is a spiritually powerful, sorcerer-like creature that practices dark magic and is therefore not to be messed with. For example, residents of northern Botswana believe that you need to move should a hamerkop land on your house, because the bird is believed to attract lightning down to earth.

The hamerkop's nest can contain 8,000 sticks and be up to 5 ft (1.5 m) wide

▲ Sturdy nest
Over 3 ft (1 m) deep and weighing up to 110 lb (50 kg), the hamerkop's gigantic nest is sturdy enough to support the weight of an adult human.

> "... the bird itself, as well as its nest, is sacred. Dire vengeance follows the man who kills it."

ROBERT GODFREY, *Bird-Lore of the Eastern Cape*, 1941

Crown feathers can be extended into a spiky crest

The triangular bill is flattened from each side, helping the bird catch frogs and fish

▲ Distinctive head
Hamerkop means "hammer head" in Afrikaans, in homage to the peculiar shape formed by its outsized head, which makes it one of Africa's most readily identifiable birds.

Order Pelecaniformes

Families 1

Species 1

Size range 18½–22 in (47–56 cm) long

Distribution Arabian Peninsula, Sub-Sarahan Africa, Madgacasar

◄ **Canopy feeding**
A black heron (*Egretta ardesiaca*) looks for fish in the shade it has created by spreading its wings over its head in an umbrellalike canopy.

► **Successful hunt**
The cryptic plumage of Eurasian bittern (*Botaurus stellaris*) is perfectly matched to its reedbed habitat, allowing it to surprise the fish that make up 80 per cent of its diet.

Herons, Egrets, and Bitterns

Ardeidae

Weighing **10 lb (4.4 kg)**, the Goliath Heron (*Ardea goliath*) is the **world's largest** heron

With their long legs, sinuous neck, and a slender, pointed bill, herons are highly specialized in capturing live prey, generally in aquatic environments.

All herons are sight hunters, either waiting motionless for prey to come within reach or stalking it through shallow water. Their specially adapted neck has an "S" bend that enables a rapid forward lunge to grab prey with their long, straight bill. Some species have brightly colored feet, which are used to startle underwater prey. Amphibians and fish are typical components of their diet, but many herons will turn their sharp bill to birds and even mammals. However, not all herons are tied to aquatic habitats. The Whistling Heron (*Syrigma sibilatrix*) inhabits savannas—from the Llanos of Colombia and Venezuela to the Pampas of Uruguay and Argentina—where it maintains a largely insectivorous diet. The Western Cattle Egret (*Bubulcus ibis*) hunts insects and frogs in open grasslands, foraging near large mammals that will flush out its prey.

World domination

The western cattle egret is also notable for its dramatic ongoing range expansion. Native to Africa and southern Iberia, in the late 1800s it crossed the Atlantic Ocean, probably from the

Order Pelecaniformes

Families 1

Species More than 60

Size range 10½-55 in (27–140 cm) long

Distribution Worldwide, except Antarctica

west coast of Africa, to northeastern South America. It went on to colonize the entire continent, reaching Tierra del Fuego by 1977, as well as the Caribbean and half of North America. After 1900, it also began a natural range expansion throughout the African continent, including Madagascar. By the 1950s, it had reached southern France and the Volga Delta on the Caspian Sea and is now an increasingly common breeding bird in England.

Wing feathers are defined by incisions and painted black

◄ Oil vessel
This heron-shaped aryballos (c.580 BCE) from ancient Greece is a clay vessel that was used to store scented oil.

Solitary hunters

Most herons and egrets are monogamous but highly gregarious, forming mixed-species roosts and nesting colonially. Breeding colonies can be very large, comprising thousands or tens of thousands of birds. Outside of these gatherings, herons typically hunt alone, which means they must move large distances between roosts and foraging sites. Even the mysterious Agami Heron (*Agamia agami*) of Central and South America, once thought to be a solitary sedentary species, has been found to form breeding colonies of more than 1,000 nests after migrating up to 745 miles (1,200 km) to join the colony.

In comparison, bitterns tend to be more solitary and territorial, even maintaining distance from each other's nest during the breeding season. Stalking through reeds, bitterns are heard rather than seen, and in times past their disembodied booming call inspired dread.

Exploited for fashion

Although several species of herons are eaten by humans, and heron flesh is often regarded as a delicacy, their plumage has been even more prized. Several groups of indigenous people

◄ Colorful yet inconspicuous
An unassuming denizen of forested creeks, the agami heron is perhaps the most graceful of all herons, with an elegantly slim neck, an extremely long bill, and spectacularly rich plumage.

> "And as a bittern bumbles in the mire, she laid her mouth down unto the water . . . "
>
> GEOFFREY CHAUCER, *The Wife of Bath's Tale*, c.1400

▲ **The woman behind the gun**
This 1911 US cartoon is a commentary on the millinery trade and the severe impact it had on birds, especially egrets. Around 500 little egrets were collected to produce a pound of silky feathers for decorating hats.

make use of heron feathers, but the scale of exploitation for Western fashion, mainly ladies' hats, during the 19th and early 20th centuries was unprecedented. Outrage in the US over the slaughter of millions of waterbirds, particularly egrets, for the millinery trade led to the foundation of the Massachusetts Audubon Society by Harriet Lawrence Hemenway and Minna B. Hall in 1896, and within two years, 16 other states had their own Audubon chapters. Their campaigning led to the passing of the Migratory Bird Act by the US Congress in 1918, outlawing the trade.

Similarly, egrets were foremost among the birds that spurred Emily Williamson in 1889 to create the Society for the Protection of Birds in the UK, which in 1904 gained a Royal Charter to become the RSPB. In 1890, the society published its first leaflet, *Destruction of Ornamental-Plumaged Birds*, to inform women of the effects of the use of feathers in fashion on the population of Little Egrets (*Egretta garzetta*) in southern Europe. By 1899, British Army officers were ordered not to use egret feathers in uniforms, but it was not until 1921 that the Importation of Plumage (Prohibition) Bill, banning the import of feathers into the UK, was passed.

BAIT FISHING

Green Herons (*Butorides virescens*) and Striated Herons (*B. striata*) are among the few birds in which tool use has been documented. Both utilize bait to attract fish to within striking range. Green herons have been recorded offering mayflies, feathers, and bread crusts while striated herons use earthworms, insects, sticks, and a variety of other items as lures.

STRIATED HERON FISHING WITH BAIT

GRAY HERON
Ardea cinerea
This large heron is found across Eurasia and sub-Saharan Africa. It feeds on fish, small mammals, birds, and even catches insects on the wing.

WESTERN CATTLE EGRET
Bubulcus ibis
This shorter-legged species prefers a drier habitat than most herons. During the breeding season, both adult males and females sport yellow plumes on the head and neck.

LEAST BITTERN
Ixobrychus exilis
Weighing just 2.8 oz (80 g), the Least Bittern is the smallest member of the heron family, and, although it can be abundant in its marsh habitat, is more often heard than seen.

Spring plumage
An adult male Little Blue Heron (*Egretta caerulea*) dressed in its spring finery steps toward the shore near Charleston, South Carolina, in this watercolor from John James Audubon's *Birds of America* (1827–1838). In the background, a mostly white juvenile, which is in the process of molting into its adult plumage, searches for prey such as fish, frogs, and shrimp.

Ibises and Spoonbills

Threskiornithidae

Admired for their often vividly colored plumage and renowned for their distinctively shaped bills, ibises and spoonbills also catch the eye with their somewhat gangly and stiff flight action.

▲ Returning to roost
The Scarlet Ibis (*Eudocimus ruber*), with its startling red plumage, is one of South America's most distinctive birds. It favors mangrove swamps, tidal mudflats, and river valleys, and roosts in huge, vivid colonies.

Waterbirds usually catch the eye, but the ibises and spoonbills are particularly memorable. Dressed in plumages that are vivid, iridescent, or adorned with elaborate plumes, these unusually shaped birds always stand out. Ibises have a long, decurved bill with a sensitive tip that is used to probe into mud or sand in search of prey and to pick up food from the ground. Spoonbills have a highly sensitive, flattened bill that broadens out at the end, which they sweep sideways through water to detect prey. Most species nest in large colonies in wetlands, often with other waterbirds, such as herons or cormorants. Some species choose to nest on cliffs, safely away from the attentions of predators, while a few species are solitary nesters. Nests are usually made of sticks with finer materials, such as grass, used as lining. Between two and five eggs are normally laid, and

Order Pelecaniformes

Families 1

Species More than 30

Size range 20–43 in (50–110 cm) long

Distribution Worldwide except for Antarctica

both adults partake in all aspects of parental care. These are generally silent birds, usually only heard giving a few grunts or croaks when on breeding grounds.

Ibises and spoonbills typically inhabit temperate or tropical climates. Although they are traditionally thought of as wetland birds, some ibises are well adapted to drier environments, such as the deserts and savannas of Africa and the high mountain plateaux of the Andes in South America. Additionally, a few species, such as the critically endangered São Tomé Ibis (*Bostrychia bocagei*), which is found only on its namesake island, inhabit dense forests, where they are seldom seen by humans.

Gods and garbage

Not only are ibises and spoonbills fascinating birds to observe, but their adaptability and resilience have intrigued humans for centuries. Perhaps the most famous example of interactions between humans and ibises comes from ancient Egypt, where the African Sacred Ibis (*Threskiornis aethiopicus*) formed an integral part of religious belief and worship. The birds would be sacrificed in their thousands every year to Thoth, a god of magic and wisdom

Different colors, such as blue and purple, appear depending on the light and the viewing angle

Shimmering plumage
The Straw-necked Ibis (*T. spinicollis*) has highly iridescent feathers that look dark in dull conditions, but a metallic glow appears in sunlight.

who is often depicted as having the body of a human, but the head and bill of an ibis. People believed that for offering the birds, which were mummified and stored in catacombs, Thoth would cure illnesses or gift long life in return. This species is now extinct in modern Egypt, although it must have once been very common along the Nile.

▲ **Clean eating**
The Australian white ibis has found a way to eat poisonous cane toads by using a stress and wash method to rid the toads of their toxins before cleaning them and finally consuming them whole.

The Australian White Ibis (*T. molucca*) is admired—sometimes begrudgingly—for its adaptability and tenacity in modern Australia, where it has adapted to live alongside humans and has been represented in both artwork and film. It is often locally referred to as the "bin chicken" due to its scavenging behavior and widespread willingness to eat garbage.

Conservation action

Both the Roseate Spoonbill (*Platalea ajaja*) in North America and the Eurasian Spoonbill (*P. leucorodia*) were historically hunted for their elaborate plumes, which were used to decorate women's hats. The latter was driven to local extinction in Britain through hunting around 300 years ago, having previously been so widespread it was a gamebird. Both species are now recovering with better protection.

The Northern Bald Ibis (*Geronticus eremita*) was hunted to extinction throughout Europe and is now the subject of reintroduction programs. In one such scheme, captive-bred birds have been shown migration routes between Austria and Tuscany, Italy, by humans using a microlight aircraft. This was inspired by the true story behind the 1996 film *Fly Away Home*, in which orphaned geese were taught how to migrate.

The **Eurasian spoonbill** was **eaten** at medieval **banquets** in Britain

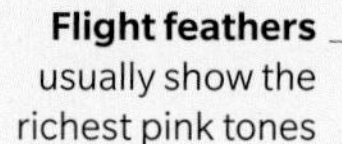

ALFRED RUSSEL WALLACE, *A Narrative of Travels on the Amazon and Rio Negro*, 1853

AFRICAN SACRED IBIS
Threskiornis aethiopicus
The African Sacred Ibis is widespread in sub-Saharan Africa and has been introduced to parts of Europe. It is one of three species with similar black-and-white plumage.

GIANT IBIS
Pseudibis gigantea
The Giant Ibis has blackish plumage with metallic silver feathers on the upperwing and is extremely rare, with only about 100 pairs remaining in Cambodia and Laos.

EURASIAN SPOONBILL
Platalea leucorodia
The largely white Eurasian Spoonbill has a yellowish band around the neck and yellow on the tip of its bill. Breeding adults have a noticeable crest.

▲ Pretty in pink
The roseate spoonbill has a largely pink plumage that is colored by carotenoid pigments in its diet.

Contemporary pressures on ibises and spoonbills primarily involve habitat loss. For example, the decline of the endangered Black-faced Spoonbill (*P. minor*) in eastern Asia was largely due to loss of its wetland habitat to industrial development. However, conservation efforts have resulted in a recent recovery in populations at some of its wintering and breeding sites.

Two of the rarest members of the family—the White-shouldered Ibis (*Pseudibis davisoni*) and the Giant Ibis (*P. gigantea*)—are beneficiaries of community-based conservation projects in Cambodia. Here, small-scale farmers use wildlife-friendly techniques to grow rice that is marketed abroad as "Ibis Rice." The scheme offers a financial incentive for sustainable farming and in turn encourages the preservation of habitat, which offers the birds a lifeline after widespread declines across their former ranges.

SENSITIVE BILL

The spoon-shaped bill, which is deep at the base but becomes flat at the tip, is swept from side to side in water to swirl up and detect favored prey, such as insect larvae, crustaceans, small fish, amphibians, and worms. The bill has many sensory cells, allowing spoonbills to hunt effectively in muddy waters by touch rather than sight.

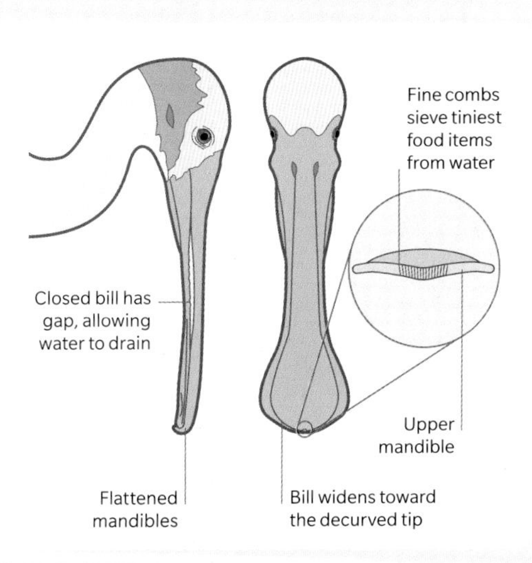

SPOONBILL HEAD IN PROFILE AND FACE ON

Hoatzin

Opisthocomidae

Order Opisthocomiformes

Families 1

Species 1

Size range 25½ in (65 cm) long

Distribution South America

Living in trees along waterways in the lowlands of tropical South America, the hoatzin is like no other bird on Earth. It is an evolutionary enigma whose exotic appearance is topped by a bizarre lifestyle.

The remarkable-looking Hoatzin (*Opisthocomus hoazin*), which clambers around waterside vegetation in tropical South America, is placed in an order all by itself. Scientists cannot agree upon its nearest relatives: cuckoos, cranes, plovers, even hummingbirds have been proposed.

The hoatzin is the only bird to have a cowlike digestive system, fermenting plant tissue in an enlarged foregut and emitting foul odors. Uniquely among tree-dwelling birds, youngsters escape predators by dropping into the water and swimming to safety. Then they use claws on their wings—a feature shared with the avian dinosaur *Archaeopteryx*—to climb back to their nest.

Hoatzins were assumed to have evolved in South America. However, recently discovered fossils suggest that their ancestor hailed from Africa or Europe. These pioneers may have drifted across over the Atlantic on an island of floating vegetation, then colonized South America. Hoatzin eggs are eaten by people in the Brazilian Amazon, and its feathers are used to make fans. Adults are hunted but for use as bait rather than to be consumed because the meat tastes as bad as the birds smell.

▲ Climbing claws
An illustration from Ogilvie-Grant's *A Handbook to the Game-Birds* (1895) shows a flightless hoatzin chick using the claws on its wings to return to the nest.

▼ Waterside perch
Three hoatzins perch on a branch in the Tambopata National Reserve, in southern Peru. They are among around 600 bird species found in the nature reserve.

Secretarybird

Sagittariidae

Towering above the African grasslands, the secretarybird is a bird of prey like no other. With its lethal long legs and huge hooked beak, it kicks its prey to death before devouring it.

Its **odd name** may be from its **crest feathers**, which resemble a row of **quill pens**

The world's most extraordinary bird of prey, the imposing Secretarybird (*Sagittarius serpentarius*), is usually found striding across the savanna in pairs or small groups. Once thought to be related to cranes, its closest relatives are hawks and eagles.

It is regarded as a beneficial bird as it kills snakes and rodents, also feeding on grasshoppers, beetles, birds, young hares, and occasionally crabs. Unusually for a bird of prey, it has short toes with blunt talons, so it cannot hold onto or carry prey, and it also searches for its food on foot, often walking long distances. Small prey may be caught with its beak and larger prey, such as snakes, can be killed with a kick, or by being stamped on and then torn apart with the beak.

During courtship a pair may soar together, uttering croaking calls, and after rising high on a thermal they may perform undulating "sky dance" displays. Although they are terrestrial birds, their nest is a structure of sticks built high in an acacia tree and it may be used for many years. A pair may roost in the nest all year or in a nearby tree. They may breed in any month of the year but will normally time it for the chick to fledge during the rainy season, when food is plentiful. Some pairs will breed again less than a month after the young become independent.

Order Accipitriformes

Families 1

Species 1

Size range 5 ft (1.5 m) tall

Distribution Sub-Saharan Africa

This young secretarybird has shorter crest feathers than adults and its upperparts are still brown

▲ Kicking its prey
Renowned for killing snakes with a kick, a secretarybird's legs are partly feathered, with thick scales lower down that may protect it from snake bites.

◄ Maturing juvenile
The bare skin on a juvenile's face changes from yellow to orange as it matures. Its black beak will become pale gray and more hooked and its eye color will change from gray to brown.

New World Vultures

Cathartidae

Weighing up to **33 lbs** (15 kg), the **Andean condor** is the **heaviest bird of prey**

These large birds of prey are conspicuous in the skies across the Americas, soaring effortlessly on their long, broad wings, from southern Canada south to Tierra del Fuego.

New World vultures are extremely capable fliers, traveling hundreds of miles without flapping their wings due to their ability to find air currents. These vultures occupy a wide range of habitats, from grasslands to rainforest and sweltering deserts to high mountain peaks, while some species, such as the Black Vulture (*Coragyps atratus*), have become successful in suburban environments as well. As the family name Cathartidae (from the Greek for "purifier") implies, these birds are scavengers, feeding primarily on carrion, which they may locate from a great distance by their suberb sense of smell. However, a black vulture will sometimes kill helpless prey, such as nestling birds or baby sea turtles.

Despite their name and appearance, New World vultures are not closely related to the Old World vultures, being genetically more similar to the secretarybird and osprey. They do not build nests, instead laying eggs on bare rock faces or in tree cavities. While condors tend to choose remote and inaccessible locations to

Order Accipitriformes

Families 1

Species 7

Size range 23–51 in (60–130 cm) long

Distribution North and South America

raise their young, other species will choose surprisingly exposed locations, with some birds even nesting under bushes on the ground.

Deep relationships

For thousands of years, these magnificent birds have been held in high esteem by various cultures. Black vultures and King Vultures (*Sarcoramphus papa*) frequently appear in Mayan codices, with the latter sometimes portrayed as a god with a human body and vulture's head that carried messages between humans and other gods. The California Condor (*Gymnogyps californianus*) has featured prominently in Indigenous American cultures. It is depicted in cave paintings and its body parts were used to make artifacts and instruments. It was also used in rituals in which birds were sacrificed, and burial sites for condors have been found. The Andean Condor (*Vultur gryphus*) is celebrated in Andean cultures, being seen as a symbol of power and health. It has been depicted in artwork dating as far back as 2500 BCE. However, some believed its body parts to have medicinal powers and this led to the killing of these birds.

◄ Emblem of the Andes
The mighty Andean condor is its namesake mountain range's most recognizable bird. Found from Colombia south to Chile, it may live for 50 years or more in the wild.

SAVED FROM EXTINCTION

By 1987, the final few remaining wild California condors—just 22 birds —were taken into captivity to start a captive-breeding scheme. This has been successful and birds have been reintroduced to several areas in the western US. Without this intervention, the species almost certainly would have been lost forever.

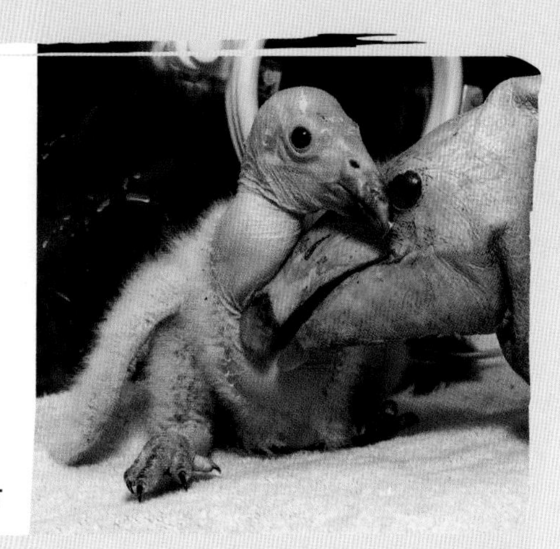

CONDOR CHICK WITH GLOVE PUPPET

> " The beauty of the California Condor lies entirely in the magnificence of its matchless soaring. "
>
> CARL B KOFORD, *The California Condor*, 1953

GREATER YELLOW-HEADED VULTURE
Cathartes melambrotus
This vulture from South America's Amazon Basin prefers undisturbed tracts of lowland forest. No nest site has ever been discovered for this species.

TURKEY VULTURE
Cathartes aura
The Turkey Vulture is common in both North and South America. It roosts communally and northern populations are migratory.

BLACK VULTURE
Coragyps atratus
The Black Vulture is the smallest member of the family, with relatively short wings. It has an obvious white patch at the wingtips.

Fleshy growths, called caruncles, protrude from the bill base—these may play a role in greeting and courtship displays

Adult bird has vivid colors on the head and neck, with one theory suggesting the brightness indicates a bird's social status

Royal vulture
The king vulture has a bald head and neck. This helps reduce the risk of bacteria becoming lodged in its feathers when it feeds on carrion.

Order Accipitriformes

Families 2

Species More than 250

Size range 9–47 in (23–120 cm) long

Distribution Worldwide

◄ **Slippery meal**
Ospreys catch fish with a spectacular swoop and swipe, gripping the slippery prey with very sharp claws and pointy spicules on the undersides of their toes.

Zeus placed Ganymede in the night sky, as the constellation Aquarius (the water bearer)

Stolen away
This 1st century CE mosaic shows the Greek god Zeus in eagle form, abducting the beautiful mortal boy Ganymede to be his cup-bearer in Olympus.

Ospreys, Hawks, Eagles, and Kites

Pandionidae, Accipitridae

Combining aerial mastery and raw power with their deadly weaponry of hooked bill and talons, the hawks and their relatives are superbly adapted top-flight predators.

Birds of prey, or raptors, attract both admiration and apprehension. All species are impressive, from the 9 in (23 cm) long Pearl Kite (*Gampsonyx swainsonii*) and Little Sparrowhawk (*Accipiter minullus*) to the $3\frac{3}{8}$ ft (1 m) tall Philippine eagle (*Pithecophaga jefferyi*) with its $6\frac{1}{2}$ ft (2 m) wingspan. Although many birds from different families will hunt other vertebrate animals, birds of prey are noted for using their taloned feet to capture and carry prey, rather than their bill. Although owls and falcons are also called birds of prey, the largest group in this category are the hawks and their relatives.

There is a long history in Europe and Asia of the use of species such as the Eurasian Goshawk (*Accipiter gentilis*) and Golden Eagle (*Aquila chrysaetos*)—as well as actual falcons—in falconry, whereby trained birds catch and immobilize prey, including quarry much bigger and heavier

Twenty-two eagle feathers form the headdress, along with numerous colored beads

The long tail is made from a further 35 eagle feathers, attached to red stroud cloth

◄ **Acts of bravery**
In this magnificent war bonnet, made by the Lakota people of the Great Plains, in the US, each of the eagle feathers represents a brave battle deed performed by the wearer.

▲ Golden hunters
Traditional Kazakh horseback hunters train and fly golden eagles, using birds taken from the wild as chicks. They are released back into the wild when aged six or seven. The birds' skills can be seen in an annual eagle festival held at Ölgii city in Mongolia.

" He watches from his mountain walls,
And like a thunderbolt he falls. "

LORD ALFRED TENNYSON, "The Eagle," 1851

than themselves. The earliest evidence of hunting with birds dates back to before 1200 BCE, in the Middle East. In 15th-century Europe, the higher class the man, the more impressive his hunting bird—a king would fly an eagle but a priest must make do with a sparrowhawk. With the invention of guns, falconry as a practical hunting method became much less widespread. In the 19th century, managed estates full of pheasants (originally imported from eastern Asia) proliferated across the UK as a popular hunting sport. To protect their gamebirds, gamekeepers turned their own guns onto the wild birds of prey, as well as on predatory mammals, and by the early 20th century many species had become very rare, or worse. The UK completely lost its breeding populations of White-tailed Eagles (*Haliaeetus albicilla*) and Eurasian goshawks, while Red Kites (*Milvus milvus*) and Western Marsh Harriers (*Circus aeruginosus*) dwindled to just a handful of pairs.

Today, birds of prey are more respected and better protected, allowing some species to recover their numbers. Reintroductions have also helped, returning the white-tailed eagle and red kite to former haunts across the British Isles. As birds of prey saw improvements in their public image, so falconry has also enjoyed a minor resurgence as a pastime. Harris's Hawk (*Parabuteo unicinctus*) is one of the most popular modern falconry birds, thanks to its agreeable nature and how easy it is to train. In the wild, this species hunts

WINGS OF DEATH

Africa's Crowned Eagle (*Stephanoaetus coronatus*) is the only wild bird species suspected of intentionally hunting humans, with a few documented attacks on young children (including the discovery of a child's skull in a nest). The Taung child, a 2.8 million-year-old fossil *Australopithecus* skull found in South Africa in 1924, was probably killed by a related species of large eagle.

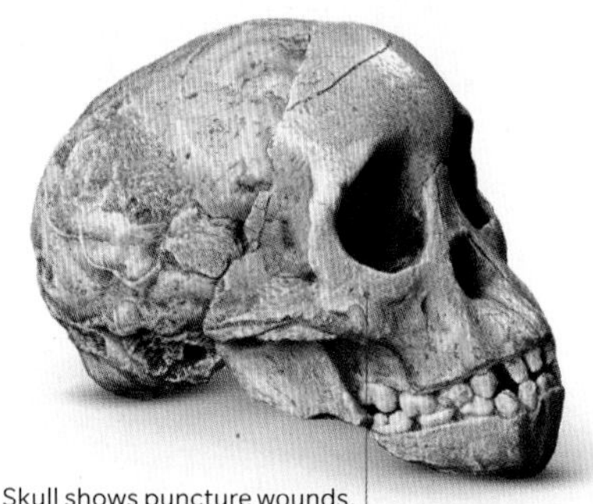

Skull shows puncture wounds, similar to those that eagles inflict on monkey prey

TAUNG CHILD FOSSIL

cooperatively in small groups, so it has a natural aptitude for forming productive partnerships. Trained Harris's hawks and their handlers now find employment as deterrents to other birds in city centers and on airfields in many parts of the world. Flying hawks regularly in these areas helps discourage feral pigeons from settling and nesting on buildings, and protects aircraft from the chance of birdstrike.

Varied diets

Birds of prey include generalist hunters but also some extreme specialists. The Osprey (*Pandion haliaetus*) feeds only on fish, swooping down to snatch them from the water's surface with a spectacular talon-swipe, while the honey buzzards (*Pernis* spp.) attack wasp and bee nests to eat the larvae. The Bat Hawk (*Macheiramphus alcinus*) is active at dusk, hunting bats as they leave their roost. The Snail Kite (*Rostrhamus sociabilis*) has an unusual long, thin hooked tip to its bill that it uses to extract snails from their shells. Plant-eating is rare, but the Palm-nut Vulture (*Gypohierax angolensis*) enjoys oil-palm fruit. Birds of prey find their food primarily by sight (hence expressions like "eagle-eyed"), although harriers, which hunt over fields in low-level flight, detect small prey moving below thanks to their sharp hearing, aided by an owl-like facial disk that funnels sound to the ear openings.

Most birds of prey are highly accomplished in the air, although they vary in their preferred flight technique. The short-winged true hawks can dash swiftly and silently through dense forest,

▲ All-consuming Sun
This Mesoamerican stone panel shows an eagle biting a cactus fruit, representing the Sun eating a human heart. Aztecs believed that such sacrifices ensured the Sun's daily return from the underworld.

➤ Symbolic salvation
North America's iconic Bald Eagle (*Haliaeetus leucocephalus*) has been rescued from a serious population decline thanks to various protections of the birds and their habitat.

Only an adult bald eagle has a white head—the color change from brown is gradual over the first few years of its life

A vivid yellow bill is typical of the fish eagles

Large eagles have a **grip strength** in their **talons** that **rivals or exceeds** the **bite force** of a **lion**

slipping through small spaces between trees to take prey by surprise. Kites wheel and hover with seemingly weightless grace, and are skilled at snatching prey and bits of carrion from the ground without needing to land. Broad-winged species such as vultures, eagles, and the hawks and buzzards from the *Buteo* genus can soar effortlessly for hours, riding to dizzying heights on rising thermals as they scan the ground below with those peerless eyes. Many Eurasian and North American cultures tell a folktale of an avian high-flying contest, although the twist in the tale is that the winner was not the eagle but the tiny songbird (commonly a wren or a goldcrest) that hid in its feathers, jumping out at the last moment to claim the prize.

Very broad wings help the vulture to soar on rising warm air currents, but make it less agile when manoeuvring

Bare feet and head are easier to keep clean for this rather messy eater

The short, wedge-shaped tail is fanned wide when soaring or when braking to land

Learning curve

For many raptor species, courtship and play both involve dramatic roller-coaster aerial chases and sometimes talon-grappling and aerial passes of food items. Juvenile fish eagles (*Haliaeetus* spp.) are particularly social and playful, developing skills that will help them best their rivals when fighting for prime position on rich feeding grounds. Endurance is as vital as agility for the long-distance migrants. Recent innovations in tracking technology have allowed scientists to follow the exact routes taken by individual migrating ospreys and European Honey Buzzards (*Pernis apivorus*), both of which cover more than 6,200 miles (10,000 km) in their southbound migrations. The data shows that the birds do their best to choose the most efficient overland route possible (sea crossings lack the crucial thermals or updrafts that help them gain and maintain height in the air). They also improve their routes through successive years, learning from mistakes made in their youth.

Gripping strength

The largest eagles are capable of killing mammals as large as foxes or young deer, and some smaller species can catch prey that is bigger and heavier than themselves. The most predatory species have tremendous grip strength, driving the

▲ Carrion feeder
Vultures of the genus *Gyps*, like this Eurasian Griffon Vulture (*G. fulvus*), are scavengers that feed on carrion and do not require great agility to catch prey.

➤ Stealthy hunter
The Eurasian goshawk is a fast, agile flier that uses stealth and agility to take prey by surprise.

PACIFIC BAZA
Aviceda subcristata
This forest raptor of Australia and Southeast Asia is unusual in that it regularly eats fruit. Bazas are nicknamed cuckoo-hawks because of their cuckoo-like barring.

ROUGH-LEGGED BUZZARD
Buteo lagopus
The species name *lagopus* means "hare-foot", referring to this raptor's feathered lower legs. It breeds across Arctic North America and Eurasia, moving south in severe winters.

RÜPPELL'S VULTURE
Gyps rueppelli
This large vulture of East Africa spends up to seven hours in the air each day and is able to fly at high altitude, despite the low oxygen levels and reduced air pressure.

long sharp talons deep into the prey's organs and thus causing death or critical injury, in a similar way to how a big cat uses its long, sharp canines and extremely powerful bite force to subdue even very large, powerful prey. Not surprisingly, other birds react strongly to the presence of a bird of prey—taking cover if the raptor is on the hunt, but going noisily on the attack if they judge that it is not an immediate threat. A sudden arrival of mobbing smaller birds can alert a birdwatcher to the presence of a bird of prey nearby. Crows and magpies are particularly aggressive, desperate to drive off the raptor before it is ready to look for its next meal. Hawks, eagles, and other birds of prey are frequently chosen as national birds and other icons, and have inspired a wealth of folk tales. The legendary and terrifying thunderbird, widespread in Indigenous North American mythology, is typically represented as a supersized condor or eagle, while actual eagles (bald and golden) are important and respected symbols of virtues including power, truth, courage, and wisdom. In Australian Indigenous folklore, the creator god Bunjil was a Wedge-tailed Eagle (*Aquila audax*)—the country's largest and most powerful bird of prey.

▲ Forked tail
An engraving from John James Audubon's *Birds of America* (1827–38) shows a Swallow-tailed Kite (*Elanoides forficatus*) killing a snake.

> " The hawk was everything I wanted to be: solitary, self-possessed, free from grief, and numb to the hurts of human life. "
>
> HELEN MACDONALD, *H is for Hawk*, 2014

Food fight
Hunger has driven this immature White-tailed Eagle (*Haliaeetus albicilla*; far right) to challenge a larger, stronger adult for the right to feed at a moose carcass lying below them in the snow on Smøla, off the west coast of Norway. The island is home to the world's densest population of these huge sea eagles.

Owls

Strigidae, Tytonidae

Order Strigiformes

Families 2

Species More than 240

Size range 5–28 in (12.5–71 cm) tall

Distribution Worldwide except Antarctica

With their expressive faces and direct gaze, owls inspire a sense of connection in humans, while their haunting calls and secretive, nocturnal ways give off an air of mystery.

Night brings the voices of owls—a quavering hoot, a breathy whistle, a sudden yelp, or an explosive hiss. Owl calls carry with startling clarity through the night air, but any sighting is usually a mere silhouette. For millennia, human societies around the world have associated these phantomlike creatures with ideas of fear, foreboding, and death.

By day, things are different. The owl becomes a bundle of soft-patterned feathers, with a sleepy but kind-eyed face, at one with the spirit of ancient forests. It is an icon of calm wisdom, companion to the wise goddesses Lakshmi in the Hindu faith and Athena of the ancient Greek pantheon. The true nature of an owl, though, is that of efficient hunter, expert hider, and devoted family member.

Sensing a hidden world

Most birds look at humans side-on, one eye at a time. An owl is different. It makes direct double-eye contact. Its face has an array of sensory tools, to find and capture prey in the dark. Forward-facing eyes provide overlapping fields of view, for accurately assessing distance. The retinas are optimized to sense contrast, and generate monochrome but exceedingly detailed visual perceptions. The owl's face shape helps direct sound to its large ear openings, which in some species, such as barn owls, sit asymmetrically to allow precision pinpointing of sound. Comblike structures along the edges of the flight feathers modify the airflow

➤ Powerful predator
One of the world's biggest owls, the Eurasian Eagle-owl (*Bubo bubo*) occurs across most of Europe and Asia. It is a formidable predator, able to kill foxes and young deer.

Feathery tufts break up the bird's silhouette to provide camouflage

Streaked or barred undersides help the owl blend in with tree bark when it is roosting

> " The clamorous owl, that nightly hoots and wonders at our quaint spirits. "
>
> WILLIAM SHAKESPEARE, *A Midsummer Night's Dream*, c.1595–96

◄ Death and mourning
Painted on a wall of the tomb of Ramesses IX in the Valley of the Kings, Egypt, owls like this were a symbol of death and mourning to ancient Egyptians. The owl hieroglyph, shown in profile with its face turned toward the viewer, represented the letter "M."

▲ Arctic cruiser
The Snowy Owl (*Bubo scandiacus*) relies on its sharp senses and physical power to detect and capture prey, such as lemmings, hares, ducks, and geese.

over the wings to reduce the sound from the wing beats. This means the owl can hear its prey more clearly, and also make a silent approach.

Habitats and hunting

A few owl species are migratory or nomadic, but most pair for life and occupy a relatively small territory. For example, a Tawny Owl (*Strix aluco*) living in a European woodland builds a detailed mental map of its "home," memorizing the best places to hunt prey. Not building nests, owls rely on preexisting places such as tree holes, which they guard fiercely. British bird photographer Eric Hosking lost an eye to a territorial tawny owl in 1937. Owls are also hostile to their fellow owls, with larger species eliminating smaller ones within their territory. This practice reduces competition for prey and nest sites, and also affects the behavior of the different species involved. The Long-eared Owl (*Asio otus*) in Ireland is widespread in all woodlands, but in Great Britain it coexists with the larger tawny owl and is, therefore, forced into marginal habitats. In mainland Europe, the tawny owl is itself excluded from some habitats by the Eurasian Eagle-owl, and fewer Tawny Owls leads to more abundant smaller owls.

The archetypal owl is a nocturnal woodland dweller and prey-pouncer, but some species live differently. The Short-eared Owl (*Asio flammeus*), found

► Ceremonial mask
This mask, carved from cedar wood by Kwakwaka'wakw artist Gilbert Dawson in the Pacific Northwest, symbolizes magic and wisdom to his people.

WESTERN BARN OWL
Tyto alba
An open-country hunter in Europe and Africa, this species is known for its ghostly appearance, acute hearing, and unearthly shrieking call.

GREAT HORNED OWL
Bubo virginianus
This mighty owl is native to the Americas. It occurs in many habitats, including gardens, forests, deserts, and swamps, and has 15 recognized subspecies.

AUSTRALIAN BOOBOOK
Ninox boobook
Australia's most widespread owl hoots "boo-book" all night during the breeding season. It feeds mostly on small mammals and reptiles, plus birds and insects.

Darkly barred flight feathers contain a melanin pigment that increases feather strength

Not all owls hoot. Some whistle, bark, hiss, screech, growl, or even make rattle calls

almost worldwide, hunts over grassland during the day. And on Genovesa, in the Galápagos Islands, it has developed a unique behavior: snatching storm petrels at their underground nests. The enormous Blakiston's Fish Owl (*Ketupa blakistoni*), found in eastern Russia and Japan, and revered as a divine messenger and protector by the Ainu people of Hokkaido, wades in rivers to grab fish and frogs with its huge feet. Found in the Americas, the charming Burrowing Owl (*Athene cunicularia*) is a real oddity. It is highly social, diurnal, lives in underground burrows, and usually pursues prey such as insects, small mammals, and reptiles, on foot, dashing along on its startlingly long legs.

➤ **Grassland burrowers**
The burrowing owl, also known by the nickname "shoco" on Aruba, lives in family groups. It either digs its own burrow or takes over the tunnels of prairie dogs and other digging mammals.

Order Trogoniformes

Families 1

Species More than 40

Size range 9–15½ in (23–40 cm) long

Distribution North America, South America, Africa, Asia

➤ Feeding time
This male resplendent quetzal is carrying a wild avocado fruit to its chicks in a nest placed in a cavity in a tree.

Trogons

Trogonidae

Trogons are stocky, long-tailed, and brightly colored birds that sit upright on branches in the forests, holding on with their unique feet.

Trogons are the only birds that have heterodactyl feet, where the first and second toes point backward and the third and fourth, forward. These forest-dwellers typically exhibit an iridescent green back, vibrantly colored underparts, and barred wings, but trogons are unobtrusive because they sit motionless for long periods, not moving unless spotting invertebrate prey. This sluggish behavior is known to the Xhosa people of southern Africa, whose name for a trogon is also slang for a lazy person.

The largest trogons are the New World quetzals. The male Resplendent Quetzal (*Pharomachrus mocinno*) was sacred to both Aztec and Mayan civilizations. Worth more than gold, quetzal plumes were arranged into a crown worn by important individuals on ceremonial occasions. In Guatemala, the currency is the quetzal.

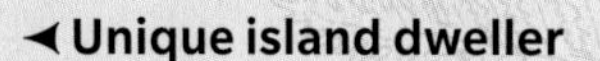

◀ Unique island dweller
A colored engraving from *The Animal Kingdom* (1828) shows a Cuban Trogon (*Priotelus temnurus*), the sole trogon on Cuba and one that occurs exclusively on this Caribbean island.

Order Bucerotiformes

Families 1

Species 3

Size range 10–11½ in (25–29 cm) long

Distribution Europe, Asia, Sub-Saharan Africa (including Madagascar)

◄ Raising the crest
As well as raising their extravagant crest upon landing, Eurasian hoopoes flaunt them in displays to demonstrate superiority over a rival or, as here, during courtship to impress a potential mate.

Hoopoes

Upupidae

This unmistakable trio of birds catch the eye with their strikingly patterned plumage and prominent crest, and also please the ear with their calls.

Hoopoes are easily recognizable thanks to their fanlike crest, their long and slightly downcurved bill, and their undulating, butterfly-like flight on strikingly patterned, piebald wings. It is perhaps surprising that many local names for these spectacular-looking birds involve onomatopoetic renditions of their song. In addition to hoopoe and the genus *Upupa*, there is *poupa* (Portugal), *poppoo* (Afghanistan), and *huh-hud* (in Arabic). The French name (*huppe*) goes further, as the adjective *huppé* now means "crested." Hoopoes are well known partly because pairs often nest in crevices in buildings or walls. Such familiarity has enshrined the Eurasian Hoopoe (*Upupa epops*) in Greek mythology. When Tereus, King of Thrace, imprisoned his wife and raped her sister, the Greek gods transformed him into a hoopoe. More positively, the Qur'an identifies the hoopoe as a messenger between King Solomon and his betrothed, and the Prophet Muhammad forbade Muslims from killing hoopoes. Ancient Egyptians considered the species sacred.

▲ Symbol of joy
A hoopoe, sometimes viewed in Chinese culture as a representation of joy, rests on a bamboo stalk in this 13th- or 14th-century painting by the Chinese artist and calligrapher Zhao Mengfu.

> " It is an adventure or an experience
> to meet them, not altogether pleasant,
> for they look exceedingly knowing. "
>
> KAREN BLIXEN, *Out of Africa*, 1937

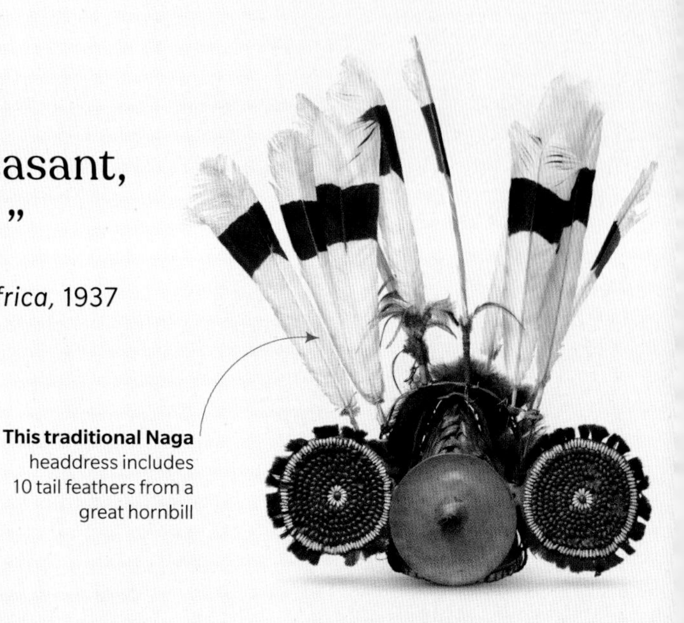

This traditional Naga headdress includes 10 tail feathers from a great hornbill

Hornbills

Bucerotidae, Bucorvidae

Few birds are so surrounded by myth and folklore as the hornbills of Africa and Asia. Their long curved bill and the unusual casque that crowns many species place these among the most striking of birds.

▲ Feather display
The Naga peoples wear headdresses decorated with hornbill feathers at festivals. Modern headdresses tend to have feathers made of paper.

In Borneo, the Rhinoceros Hornbill (*Buceros rhinoceros*) is known as the king of the birds. Among the Naga communities in India and Myanmar, the feathers of the Great Hornbill (*Buceros bicornis*) are a symbol of loyalty and devotion, worn during festivals and displays. One Naga story tells of a boy who ran off with a girl from another village. Trapped in a tree where he was collecting fruit, he slowly turned into a hornbill. One of his feathers fell on the lap of his heartbroken lover, who treasured the feather ever after. In Africa, meanwhile, the Southern Ground Hornbill (*Bucorvus leadbeateri*) is associated with rain—the sighting of one in Malawi means it is time to prepare the fields.

Hornbills range in size from medium to very large. Ground hornbills are especially heavy and walk on open ground, while other large hornbill species live in dense forests and perch high in trees. The long, slender-tipped bill is used dexterously to reach for fruit or catch large insects and small rodents, which are swallowed with a backward jerk of the head. Many species of hornbills have an unusual horny protuberance above the bill called a casque. These peculiar structures vary in color, shape, and size, and are usually almost hollow.

A sealed deal

Most species form strict pairs and nest inside a cavity in a tree or in tumbled rocks. To keep out predators, the pair reduce the entrance to a tight fit with mud and droppings before the female enters. The male adds further material to seal his mate inside for several days before she lays the eggs, leaving a slit through which he can provide food. The female eventually breaks out when space becomes too tight.

► Feeding time
A pair of Southern Yellow-billed Hornbills (*Tockus leucomelas*) feed their chicks. Having been sealed in the nest hole, the female emerges when the chicks grow larger.

The male hornbill does not enter the nest but provides food to the female and chicks throughout the nesting period

Order Bucerotiformes

Families 2

Species More than 60

Size range 13–71 in (32–180 cm) long

Distribution Africa, Asia

RHINOCEROS HORNBILL
Buceros rhinoceros
A forest hornbill that feeds on fruit, this is Malaysia's national bird. Bright colors are added to the casque by oils spread from the preen gland above the bird's tail.

Female hornbills remain **self-imprisoned** in their nest hole for more than **three months**

SOUTHERN GROUND HORNBILL
Bucorvus leadbeateri
While it feeds almost entirely on the ground, this massive bird nests high in trees. Its short flights reveal striking white wingtips. Ground hornbills may live for more than 30 years.

⅄ Unusual beauty
The great hornbill is widespread in southeast Asia and parts of India. It raises its head and bill in courtship, but the casque has no obvious function; though it may help in attracting a mate.

Kingfishers

Alcedinidae

Order Coraciiformes

Families 1

Species More than 110

Size range 4–18 in (10–46 cm) long

Distribution Worldwide except Antarctica

One of the most colorful bird families, kingfishers have a large head compared to their body and a distinctive long bill. Despite their name, many are land birds that do not eat fish, but all are effective predators.

Pied kingfishers may beat fish around 100 times before swallowing

Strong wings allow the bird to hover over water, hunting across large lakes, and even at sea, without perching

A large daggerlike beak is used to catch fish up to 5 in (13 cm) long

◄ In-flight feeding
Pied kingfishers can manipulate and swallow their prey during flight. They have been known to catch two fish in one dive.

This cosmopolitan family is found in many different habitats, ranging from high mountains to coral atolls. While many kingfishers live along coasts, beside rivers, and in wetlands, more than half are found in woodlands, and some even in deserts. One of the world's most numerous and familiar kingfishers is the Common Kingfisher (*Alcedo atthis*), which ranges across Europe, Asia, and northern Africa, and into Australasia. By contrast, the long-tailed Kofiau Paradise Kingfisher (*Tanysiptera ellioti*) has a tiny distribution area, restricted to the Indonesian island of Kofiau.

Kingfishers do not have a great vocal range, producing mainly trills, rattles, and whistles, but they can be very noisy—most famously, the dawn and dusk choruses of Australia's Laughing Kookaburra (*Dacelo novaeguineae*). Most, but not all, kingfishers have three forward-pointing toes, with the outer and middle toes fused at the base (a condition known as syndactyly).

Halcyon days

The *Ceyx* genus of kingfishers derives its name from Greek mythology. Legend tells how Zeus was angered by the relationship between Princess Alcyone and King Ceyx, and when Zeus killed Ceyx, Alcyone threw herself into the sea. The pair were then reunited when they were both turned into kingfishers. (Another version has Ceyx lost at sea and Alcyone taking her life on hearing the news.) The term "halcyon days" (a period of peace and happiness) may also derive from this myth, as Alcyone's father is said to have calmed the seas for a week before and after the winter solstice in December so that his daughter could safely lay her eggs in a nest on the beach. In Polynesia, the Sacred Kingfisher (*Todiramphus sanctus*) was so called because it was seen as a holy bird and was worshipped for its believed power over wind and water.

In the UK, during the Victorian era, common kingfishers were hunted for their beautiful feathers. The birds were stuffed and displayed in glass cases, and their feathers used to decorate hats, jewelry, and other fashion items. In China, the iridescent blue feathers of kingfishers have been used for the past 2,000 years as an inlay for objects such as hairpins, headdresses, and fans. The feathers are carefully glued onto gilt silver to produce tian-tsui, a decorative process that is similar to cloisonné. Common kingfisher

◄ Contrasting colors
The Banded Kingfisher (*Lacedo pulchella*) is found only in Southeast Asia. It has highly distinctive sexual dimorphism: males are bright blue and black with an orange (or black) face, while females have chestnut and black plumage.

▼ Delicate design
This hand-painted 18th-century Japanese inrō (a traditional lacquered case used for carrying small objects) has a decorative ceramic netsuke to attach it to an obi (sash).

A common kingfisher perches on the edge of a boat looking for prey

Maintenance routine

Native to North America, the Belted Kingfisher (*Megaceryle alcyon*) preens its feathers to remove dirt and maintain waterproofing after divinginto water. It uses its bill to smear preen oil over its plumage.

Preen oil is collected from a gland at the base of the tail

feathers contain brown pigments, and their blue color comes from a phenomenon known as "structural coloration" (first described in the 17th century by British scientist Robert Hooke), in which reflected light makes the feathers appear blue. Many other kingfisher species have similar semi-iridescent blue plumage.

When hunting, kingfishers are prepared to wait, sitting motionless on a perch, carefully watching a stream or forest floor. They have favorite perches, and return to them again and again. They rarely chase their prey, preferring to suddenly drop onto it at speed. The Hook-billed Kingfisher (*Melidora macrorrhina*) digs into soil and leaf litter for invertebrates, while the Shovel-billed Kookaburra (*Dacelo rex*) uses its large wide bill as a shovel to plow through soil or mud to look for worms, insects, and snails.

River kingfishers (*Alcedo, Ceryle, Megaceryle,* and *Chloroceryle* spp.) feed mainly on fish and other aquatic invertebrates. They have a nictitating membrane (third eyelid) that covers the eye when they hit the water. Uniquely, the Pied Kingfisher (*Ceryle rudis*) has a bony plate, which slides across to protect the eye. Terrestrial kingfishers like to feed on big insects, and the largest species will also take reptiles, amphibians, mammals, and even birds. Larger prey is often taken to a perch and beaten against it to kill it and break off spines. Fish are generally manipulated so they can be easily swallowed head first, and pellets of indigestible bones and

GIANT KINGFISHER
Megaceryle maxima

At up to 18 in (46 cm) long and weighing as much as 15 oz (425 g), this species is the largest of the African kingfishers.

WHITE-THROATED KINGFISHER
Halcyon smyrnensis

The white throat and contrasting chestnut brown and brilliant blue plumage make this Asian kingfisher species easy to identify.

LAUGHING KOOKABURRA
Dacelo novaeguineae

Native to Australia, this species is found in woodland, farmland, parks, and gardens. Its iconic "laugh" is a territorial call.

scales are then coughed up. The laughing kookaburra catches snakes with its powerful hook-tipped beak, sometimes dropping them from a height or smashing them against a rock before swallowing them head first. The Great-billed Kingfisher (*Pelargopsis melanorhyncha*) eats mainly crabs, while the Red-backed Kingfisher (*Todiramphus pyrrhopygius*) breaks into martins' nests to eat their chicks.

Nesting in holes

All kingfishers nest in holes, and like most cavity nesters, they lay white eggs. Tree-nesting birds usually take over a nest from a different species, such as a woodpecker, and will only construct their own nest in soft rotting wood. Others build in the earthen tree nests of termites. Earth or sand banks are also used as nesting sites, and kingfishers dig out long tunnels that slope downward from the egg chamber to the opening, which is always sufficiently high above water to avoid flooding. The Giant Kingfisher (*Megaceryle maxima*) has been known to build a tunnel 28 ft (8.5 m) long.

Most kingfishers are monogamous, and both parents help with nest building and caring for the young. Many kingfishers are solitary outside the breeding season, but the pied kingfisher and laughing kookaburra usually live in family groups, and a breeding pair may have one or more helpers. Pied kingfishers are even known to form roosts of up to 220 birds.

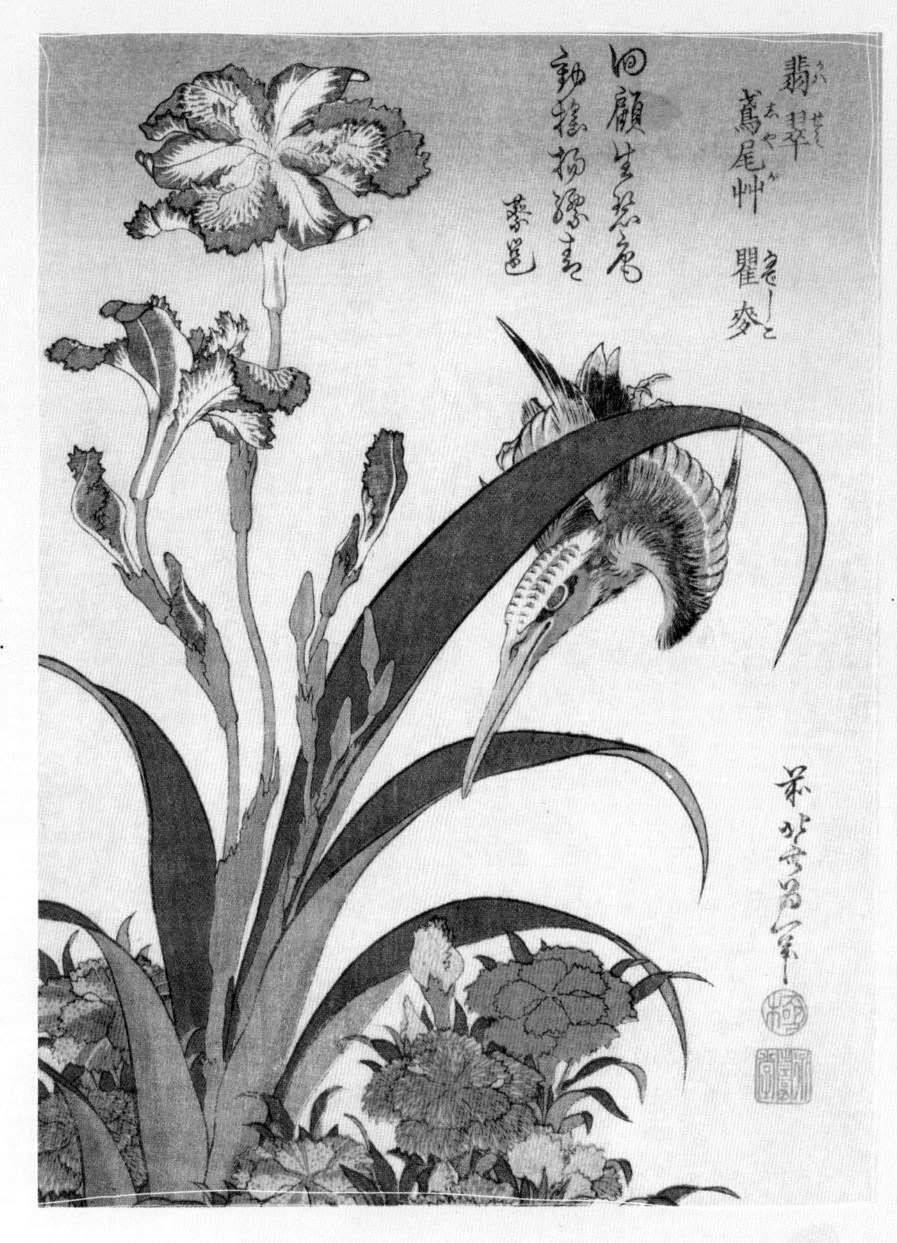

▲ Eyes on the prize
This c.1834 Japanese woodblock print by Katsushika Hokusai depicts a common kingfisher among irises and wild pinks. The bird appears to be making a rapid descent, probably honing in on its prey.

TARGETING PREY IN WATER

Kingfishers have a special sensitive area in their retina that allows them to accurately target prey in water. They have a second fovea (an area with a high density of cones and where vision is sharpest), separate and to the side of the main one, and they use this when entering the water. It gives them monocular vision in the air and binocular vision in water, which helps them to judge the precise location of moving prey and plan their dive.

ADJUSTING ANGLE OF ATTACK

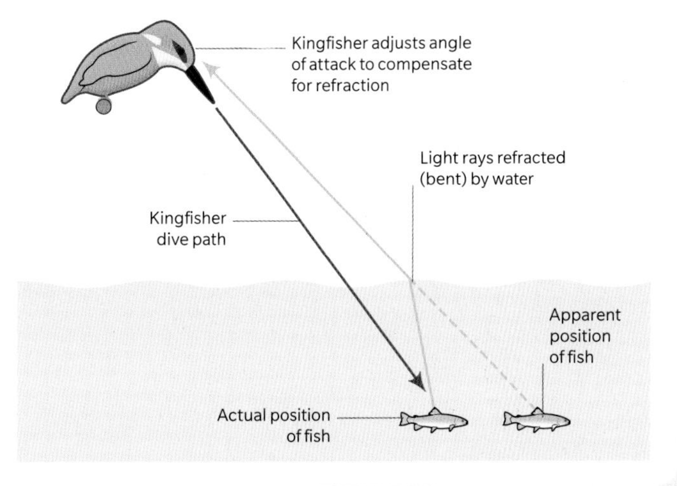

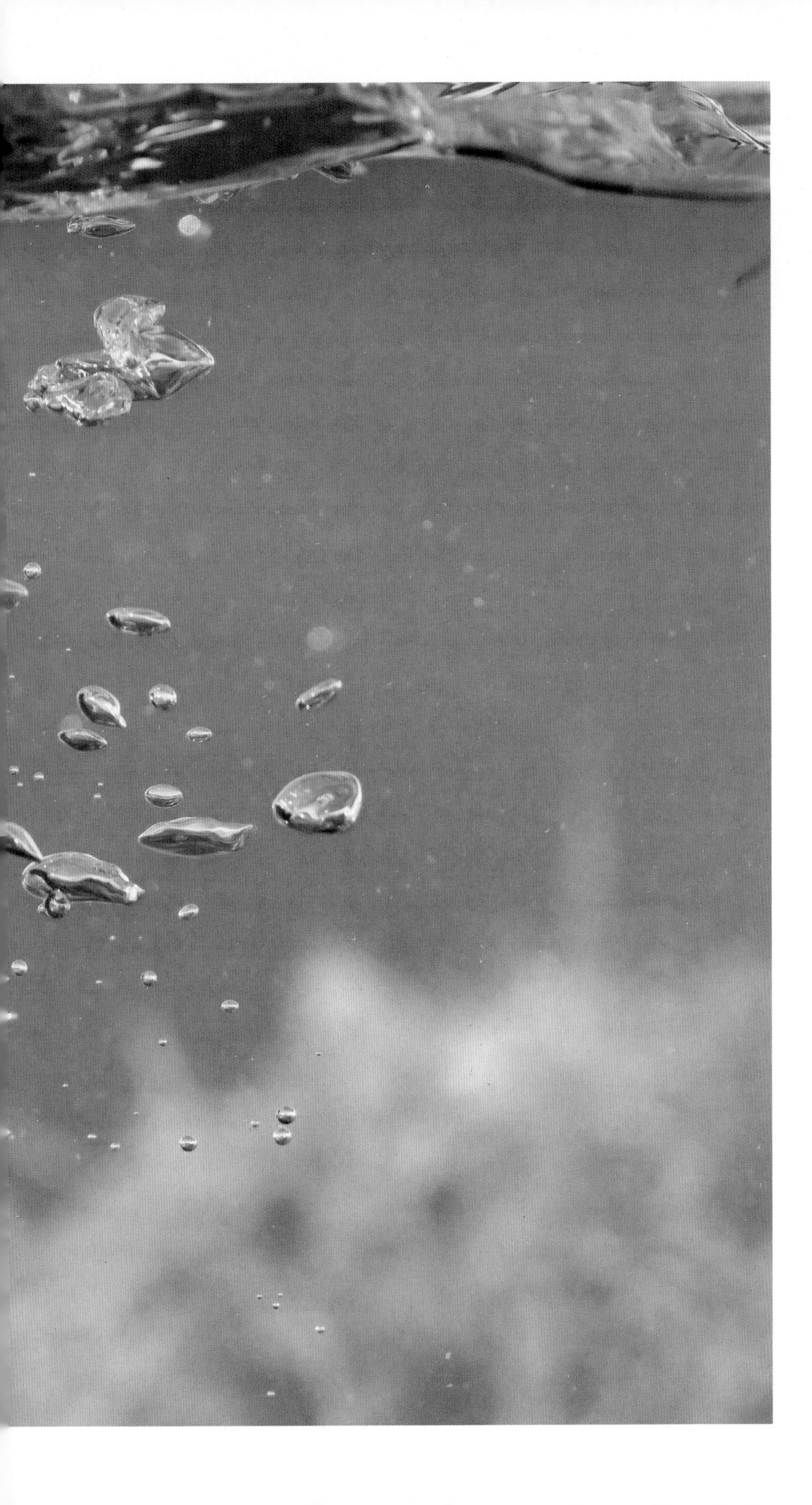

Right on target
Waterproof plumage and hollow bones make the common kingfisher so buoyant that it cannot stay submerged long enough to chase down prey. Instead, it targets a fish from a favorite perch above the water then dives straight at it. Having seized the fish with its bill, it flaps its wings a few times and buoyancy takes it back to the surface.

◄ Colonial nesters
Like many other members of the family, Red-throated Bee-eaters (*Merops bulocki*) breed together in large colonies. They remain in the nesting area year round.

► Aerial master
The European Bee-eater (*M. apiaster*) is the most widespread of its family. It is migratory, breeding across Eurasia from Iberia to Mongolia and wintering in sub-Saharan Africa.

Bee-eaters and Rollers

Meropidae, Coraciidae

Their plumages awash with vivid colors, these are confident and often conspicuous birds of open habitats or forest clearings. They have captured the imaginations of people for millennia.

Order Coraciiformes

Families 2

Species More than 40

Size range 6–16 in (15–40 cm) long

Distribution Eurasia, Africa, Australasia

An encounter with a bee-eater or roller might typically begin with the sight of one perched motionless on a prominent dead branch or overhead wire. Then, in a sudden burst of energy, the bird is up and away in pursuit of a meal, revealing its exquisite plumage adorned in vivid hues, often of blue or green.

Few birds can match these for their allure. As well as species with beautiful plumage, bee-eaters and rollers also include birds with exceptional aerial skills—they are a joy to watch as they hunt or interact on the wing. Indeed, rollers are named for their acrobatic flight displays, in which they drop from a height with their wings and body twisting and turning in spectacular fashion.

Setting up home

Both bee-eaters and rollers are cavity-nesting birds, but they have very different nesting habits. Bee-eaters are primary cavity nesters, which means they dig their own burrows in banks of earth. Rollers, meanwhile, are secondary cavity nesters, meaning rather than constructing their own nests, they instead use natural tree cavities, rock crevices, or holes excavated by other birds.

Just one European bee-eater may eat as many as 250 bees in a single day

▲ Snug roosters
Little Bee-eaters (*Merops pusillus*) will huddle together on a branch to spend the night. As well as warmth, this offers them safety in numbers.

Most species of bee-eaters are highly social—birds may form huge colonies that sometimes include hundreds of pairs. The relationships between the birds are complex, and they sometimes participate in cooperative breeding, with related adults helping a nesting pair by feeding the young. Between two and seven eggs are laid, and studies have shown that recruiting these helper adults increases the number of young fledging the nest.

Rollers, meanwhile, are much more independent. They are solitary nesters, laying between two and four eggs. This can leave nests at greater risk from predators, but rollers have developed some remarkable techniques to deal with them. For example, when threatened, European Roller (*Coracias garrulus*) chicks vomit a foul-smelling orange liquid on themselves as a defense mechanism against snakes and birds of prey. It also makes them a far less attractive meal.

> The lilac-breasted roller mates for life—it is associated with love and marriage in southern Africa

PERCH THEN POUNCE

Both bee-eaters and rollers make effective use of the "sit and wait" hunting strategy. The bird sits on an exposed perch (or, in the case of this lilac-breasted roller, the back of a zebra) and waits there attentively. As soon as it spots a potential prey, it swoops down and grabs it with its bill.

WATCHING FOR PREY

A place in the sun

Bee-eaters and rollers are usually found in open country, favoring savanna or woodland—only a few bee-eater species inhabit closed-canopy forests, where they live inconspicuous lives in the understory. They are warmth-loving birds, with most species favoring tropical or subtropical climates year-round, while those breeding in temperate regions migrate toward the equator during winter. Most species have a favorable conservation status, with only a few showing signs of decline due to habitat loss.

Bee-eaters prey on a wide variety of flying insects, ranging from tiny flies to dragonflies. Many of these can easily be consumed on the wing, but their favored food of bees and wasps requires greater care. To avoid being stung, the birds return to a perch where they then beat and rub the prey against a branch to remove both the sting and the venom before eating.

Their association with bees has been known for centuries. In the 4th century BCE, Aristotle, the ancient Greek philosopher, wrote in his *History of Animals* that the feeding habits of European Bee-eaters (*Merops apiaster*) posed a threat to bee-keepers. He describes how they destroyed the birds' nests to protect the insects.

Rollers are bigger and more robust than bee-eaters, and have strong bills. They can therefore prey on larger animals. Although they will catch prey on the wing, in many cases rollers will pounce on their prey on the ground. Their choice of prey varies from locusts, scorpions, and frogs to lizards and voles. Sometimes, rollers will even prey on weakened small birds.

The strong bill is adapted to grasping larger prey

◄ High roller
Both male and female lilac-breasted rollers have the same brightly colored plumage and perform acrobatic displays.

The tail streamers may act as rudders, helping the bird's flight performance

Worldwide appreciation

It is no surprise that such striking birds have long been appreciated by people around the world. Indian Rollers (*C. benghalensis*) feature prominently in Hindu myths and are associated with the god Shiva. The birds used to be caught, their feathers taken and chopped to feed to cows, because it was believed that this would improve milk yield. Today, it is the state bird of three Indian states. The Lilac-breasted Roller (*C. caudatus*) is unofficially the national bird of Kenya.

Despite its religious significance, hunters in the early 20th century posed a serious threat to the Indian roller. It was killed in large numbers for its colorful feathers, which were sold by traders to be be used to decorate dresses, hats, and other clothing accessories.

In ancient Egypt, bee-eaters were considered to have medicinal properties. Their fat was prescribed to deter biting flies, while an unspecified ailment was treated by applying smoke from charred bee-eater legs to the eyes. In Hinduism, the distinctive profile of bee-eaters in flight was said to resemble a bow and arrow, leading to a Sanskrit name meaning "Vishnu's bow" and a close association with archery.

ORIENTAL DOLLARBIRD
Eurystomus orientalis
Named for the coin-shaped spot on each underwing, this roller is widespread in Southeast Asia. Some birds migrate to Russia, China, or Australia to breed.

► Renaissance study
Most rollers have vibrant azure flight feathers, as captured here in immaculate detail in Albrecht Dürer's anatomical study of a European roller's wing from 1512.

RED-BEARDED BEE-EATER
Nyctyornis amictus
This large bee-eater is resident in lowland forests across Southeast Asia. Its plumage is green with a red face and breast, and it has a startling orange eye.

Motmots

Momotidae

Although they boast colorful plumage and an extravagant tail, motmots sit motionless in the understory of dark forests so are tricky to spot.

Motmots are the sole bird family whose center of diversity lies in Central America, where ten species live. Ten species have a long tail with unusual racquet- or spatula-like tips on the central pair of feathers. When the first European account of motmots—from 16th-century Spanish explorer Francisco Hernández—highlighted the tail, English ornithologist John Ray was sceptical that the bird was real.

Initially, ornithologists suggested that forest-dwelling motmots snipped away at their tails deliberately, for decorative purposes. A more elaborate explanation came from the Pareci people of Mato Grosso (Brazil), who told how the flame-toned Rufous Motmot (*Baryphthengus martii*) helped their ancestors acquire fire by carrying an ember on its tail, which burned away the feathers. The reality seems to be that weaker parts of the feathers simply wear away.

▲ Forest predator
The Broad-billed Motmot (*Electron platyrhynchum*) is a consummate understory predator, swooping on large insects, such as this moth, and seizing them in its serrated bill.

What's in a name?

The word motmot probably comes from the Nahuatl language spoken by the Aztecs, and is thought to be a description of one species' hunting behavior. Local names in Guyana and French Guiana (respectively *houtou* and *motmot houtouc*) also reflect a typical motmot's deep, almost owl-like vocalizations, usually uttered from the rainforest understory. In Brazil, one local name of the Rufous-capped Motmot (*B. ruficapillus*)—*passáro-péndulo* (pendulum-bird)—has a very different origin. Particularly when anxious, motmots often wag their tail slowly and unpredictably back and forth, or side to side, or even frozen at an unexpected angle.

Order Coraciiformes

Families 1

Species More than 10

Size range 6½–19¼ in (16.5–49 cm) long

Distribution Mexico, Central America, South America

The long tail has unusual spatula-shaped tips to the two central feathers

▲ Taxonomy lesson
Lesson's Motmot (*Momotus lessonii*) is one of five species that, until 2009, were all considered to be forms of the blue-crowned motmot.

> "What is rare and extraordinary in this Bird is, that it hath in its Tail one quill longer than the rest."
>
> FRANCISCO HERNÁNDEZ, *A history of the birds of New Spain*, 1651

Order Piciformes

Families 1

Species More than 10

Size range 4–8 in (10–20 cm) long

Distribution Africa, Asia

Honeyguides have **thickened skin**, thought to **protect** them from **bee stings**

Honeyguides

Indicatoridae

▲ **Wax-eater**
A Lesser Honeyguide (*Indicator minor*) clings to an exposed wax comb, which it is able to digest due to special enzymes in its pancreas and small intestine.

Honeyguides derive their common name from the unique behavior of just one species, which leads people to bee nests in return for access to wax and bee larvae.

In the 16th century, a missionary in present-day Mozambique noticed a bird entering his church to peck wax from candlesticks. He reported that, oddly, the self-same species, which became known as the Greater Honeyguide (*Indicator indicator*), led people to bee nests by calling and flying between trees. After the men harvested the honey, the bird would eat the discarded wax combs plus bee grubs. Honey-hunters use a specialized sound that encourages the birds to seek hives—a receptiveness to human instruction hitherto associated only with domestic animals, such as dogs.

All honeyguides can smell beeswax, which helps them find bee nests, and they also exhibit other interesting behaviors. Like cuckoos, they are brood parasites, laying their eggs in another species' nest. The female honeyguide may puncture the host's eggs to ensure that her chick is the sole beneficiary of food brought by the unwitting foster parents. Any host chick that does manage to hatch is swiftly killed by the honeyguide chick using a sharp hook on its beak.

➤ **Raiding a hive**
An engraving from *The Naturalist's Pocket Magazine* (1800) shows a greater honeyguide feeding on beeswax from a hive.

Barbets

Capitonidae, Lybiidae, Megalaimidae

Order Piciformes

Families 3

Species More than 90

Size range 3½–12 in (9–30 cm) long

Distribution South America, Africa, Asia

Related to woodpeckers, with similarly strong feet but heavier bills, barbets are as likely to be seen on the ground, or clambering about a termite mound, as in a tree.

Barbets form three distinct geographical groups: in South America, Africa, and Asia, where they range from the Himalaya to Indonesia. All species are large-headed and short-tailed. The larger barbets look a little like smaller-billed, colorful toucans and are mostly very vocal, but often difficult to see in the forest foliage. Smaller African species, known as tinkerbirds, are like stocky, thick-billed tits, and are more elusive in the tree canopy.

Barbet calls are mostly repetitive, echoing notes, or a series of trills that can last for several minutes. These give rise to several names such as Coppersmith Barbet (*Psilopogon haemacephalus*). Some African barbets are locally known as "fruit salad birds," referring perhaps to their mixed diet, with as many as 60 different fruits being eaten, or more particularly the spotted yellows, reds, black, and white of their plumages.

▼ Living at altitude
The Great Barbet (*P. virens*) is found in high-altitude forests in Asia, from India and Nepal to southeast China, where it eats fruit and insects.

Barbet pairs may **sit** on a **tree top calling** in a long and loud **duet**

The dark olive back and black head make the bird hard to spot in dense forest

The broad, heavy but fine-tipped bill is used to excavate nest holes and pick up insects

Savanna barbets
D'Arnaud's Barbet (*T. darnaudii*), seen here in an illustration by 19th-century Dutch artist JG Keulemans, is a small species that is common in African savannas.

RED-HEADED BARBET
Eubucco bourcierii

A New World barbet found in mountains from Central America south to Peru, this species creates a characteristic forest sound with its frequent, prolonged, and resonant trilling calls.

YELLOW-RUMPED TINKERBIRD
Pogoniulus bilineatus

This small African woodland species is striped with white on the the face, which separates it from similar tinkerbirds. It has repetitive calls or trills, depending on its geographical location.

Most species remain little studied, but barbets are known to be monogamous. Breeding pairs share the duty of incubating up to seven eggs for around 15 days, and the chicks fledge after three or four weeks. Several species, including the Red-and-yellow Barbet (*Trachyphonus erythrocephalus*), are known to form communal nests. Barbets that live in forests require soft or dead and decaying wood in which to excavate their nest holes, whereas species living in more open habitats use termite mounds or firm earth.

Closer to toucans

The 15 species of South American barbets are more closely related to the toucans than to the other barbets. The Scarlet-banded Barbet (*Capito wallacei*) was discovered as recently as 1996 and is known only from one forested ridge in Peru, with an estimated population of fewer than 1,000 individual birds. It was thought to have two very similar forms but, with the aid of DNA studies, one of the forms has since been elevated to a separate species, now called the Sira Barbet (*C. fitzpatricki*).

The black crest is used in display but stiff, shieldlike facial feathering helps protect against biting insects

▲ Ground hopper
Often seen on the ground, the common and colorful Crested Barbet (*Trachyphonus vaillantii*) typifies the active nature of the African barbets with its jerky hops and short flights.

BROWN-HEADED BARBET
Psilopogon zeylanicus

A rainforest bird in southern Asia, the Brown-headed Barbet has also moved into wooded suburban areas such as parks and large gardens. This species is well known for its repeated calls.

Toucans

Ramphastidae

Toucans are colorful, tree-dwelling, mainly fruit-eating birds with enormous bills. Many are ambassador species for threatened rainforests and cloud forests across the Americas.

Order Piciformes

Families 1

Species More than 40

Size range 12–24 in (30–61 cm) long

Distribution Mexico, Central America, South America

▲ Rainbow colors
The keel-billed toucan's strikingly patterned bill can measure up to half the bird's total body length, making it one of the largest beaks relative to body size of any bird.

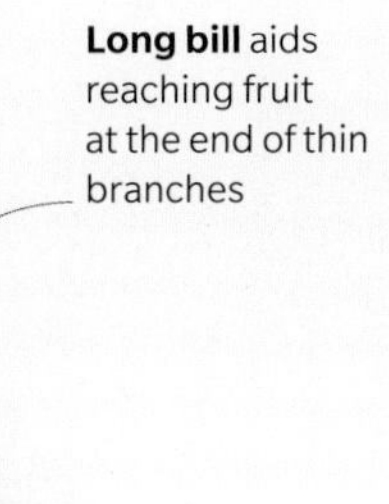

LIGHTWEIGHT BILL

Although toucan bills look heavy, they are actually hollow and thus very light. Bony struts are organized crosswise to support the outer sheath. The bill helps keep the bird cool under the hot tropical sun. Toucans are able to flush blood through a network of vessels in their bill, and its large surface area sheds body heat efficiently. In cool spells, the blood flow is restricted.

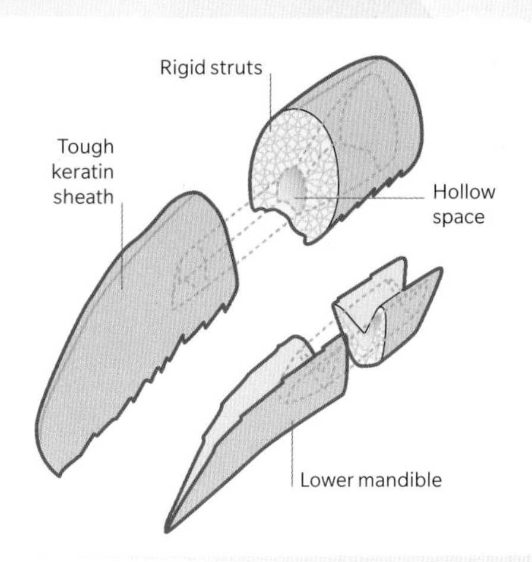

BILL STRUCTURE

HOODED MOUNTAIN TOUCAN
Andigena cucullata
There are four species of mountain toucan, and they all live in montane forests, specifically in South America's Andes, from Peru into Bolivia.

Most toucans inhabit forests, from humid lowland rainforests in the Amazon basin to mist-bound montane cloud forests in the Andes. They are social birds, feeding in small, restless groups. Some araçaris (*Pteroglossus* spp.) breed cooperatively, and offspring from previous years help their parents raise the subsequent brood.

In artwork and advertising, as in life, these colorful birds with their preposterous beaks are instantly recognizable. Known locally as the "bill bird," the Keel-billed Toucan (*Ramphastos sulfuratus*) became the national bird of Belize in 1981, when the nation achieved independence. In 1935, the Toco Toucan (*R. toco*) began its reign as a mainstay of Guinness (Irish stout beer) adverts.

Curiosity killed the toucan

Sadly, the popularity of toucans has often been to their detriment. Their innate curiosity—they can be attracted by imitations of their calls—renders them simple to capture or kill. In Brazil, toucans have been hunted for their tasty meat, at least into the late 20th century, and their bills ground into an allegedly medicinal powder. Unsurprisingly, the value of toucans was also linked to their rainbow hues. In the 18th and 19th centuries, Brazilian monarchs Pedro I and Pedro II both wore a ceremonial cape made of yellow breast feathers from Channel-billed Toucans (*R. vitellinus*). And in some cultures, such as the Jivaro people of Peru, enormous extravagantly colored toucan bills were worn like jewelry.

▲ Iconic brand association
The toco toucan, as drawn by John Gilroy, was the most celebrated of the animals in the Guinness adverts for nearly 50 years.

EMERALD TOUCANET
Aulacorhynchus prasinus
No toucan occurs farther north than the emerald toucanet. Although brightly colored, its green tones offer surprisingly good concealment in the rainforest canopy.

CURL-CRESTED ARAÇARI
Pteroglossus beauharnaisii
Probably the strangest-looking toucan, this species has modified head feathers that resemble shiny black plastic, structured into what look like curls.

> "Of the many excesses in the bird world, none seem quite so illogical as the large, colorful, and flamboyant beaks of toucans and toucanettes."
>
> HENRY BATES AND ROBERT BUSENBARK, *Finches and Soft-billed Birds*, 1970

◄ Braced for action
India's Black-rumped Flameback (*Dinopium benghalense*) uses its stiff tail as a stabilizing prop, which is typical of most woodpeckers. Combined with sharp curved claws, this creates a firm foundation for feeding.

Woodpeckers

Picidae

Order Piciformes

Families 1

Species More than 230

Size range 3–21½ in (7.5–55 cm) long

Distribution North and South America, Europe, Africa, South and Southeast Asia

Clinging to trees with sharp claws and a fierce grip, woodpeckers are able to chip and hammer at bark to extract insect larvae, open nuts, or allow sweet sap to ooze out and attract prey.

Woodpeckers are extraordinary birds, not least because there are some strikingly patterned species. When a Great Spotted Woodpecker (*Dendrocopos major*) arrives in a dramatic whirr of wings on a branch or garden bird feeder, for example, it creates an eye-catching swirl of black, white, and scarlet.

Within the woodpecker family, colors and markings follow several basic patterns. Many species have distinctive crown or nape patches, sometimes combined with a broad "mustache," which is usually more brightly colored in males. These patches are often a vivid red or orange color. The back feathers may be bright yellow-green, brown, or piebald, often neatly barred with black and white to give a silvered effect.

Most woodpecker species are heard more often than they are seen. Many have striking calls and cackles, which develop into simple but ringing, laughing songs in spring. Traditionally, in parts of Europe, woodpecker calls were thought to herald the onset of rain, while North American legends associate them with friendship and happiness.

Distinctive drumming

Some woodpecker species make "drumming" sounds instead. Performing all the functions of song, drumming is a proclamation of a woodpecker's presence and territorial claim in spring, often carrying great distances through woodland. It is a rapid rattle, created simply by drumming the tip of the bill

► Engraved shell
This shell gorget (a type of pendant, often worn as a status symbol) is from the Mississippian culture and probably dates to c.1250-1450 CE. Engraving the hard shell was skilled and time-consuming labor.

◄ Family gatherings
A pair of Hairy Woodpeckers (*Leuconotopicus villosus*), top left, and Red-bellied Woodpeckers (*Melanerpes carolinus*), top right, the male with its tongue extended, are shown with other members of the woodpecker family in John J. Audubon's *Birds of America* (1827–38).

against a resonant branch or pole to give a short burst of sound. The characteristics of the drumming are distinctive for each species. While the Lesser Spotted Woodpecker (*Dryobates minor*) makes a soft relaxed drum, the larger Great Spotted Woodpecker (*Dendrocopas major*) has a shorter, more abrupt, echoing burst. Females have a weaker drum, which can make identification complicated. In Norse mythology, woodpeckers are linked with Thor, god of thunder; in ancient Rome, they were associated with Mars, god of war, and also with the fertilization of soil.

Subspecies are split into red- and yellow-shafted forms

Vivid statement
The conspicuous bright underwing and tail identify this bird as a yellow-shafted form of North America's northern flicker.

Chipping away

Most woodpeckers are forest-dwelling birds and feed in trees or around fallen stumps. Muscles and cartilage in the neck and at the base of the bill, a spongy skull, and a tightly fitted brain help them survive harsh blows as they chip into tree bark to feed. These anatomical features also enable them to hammer rapidly to produce their drumming sounds, or chisel energetically to excavate nest holes. Woodpecker holes are important in the ecology of forests. Many other bird species, such as the Common Starling (*Sturnus vulgaris*) and some chickadees and owls, are unable to make their own nest holes, so they occupy old woodpecker holes.

Not all woodpeckers hammer into trees to find food. Some, such as the European Green Woodpecker (*Picus viridis*), concentrate on ground feeding, searching for ants. If disturbed, they fly upward in a deep swooping undulation (a typical woodpecker action), created by a sudden burst of wingbeats. Alternatively, they bound away to hide behind a tree trunk, often peering around the side to assess possible danger.

Lost forever?

Two particularly large woodpecker species are perilously close to being declared extinct. One is the Ivory-billed Woodpecker (*Campephilus*

YELLOW-BELLIED SAPSUCKER
Sphyrapicus varius
Eastern North America's only fully migrant woodpecker spends the winter in the Southeastern US, Central America, and the Caribbean.

AFRICAN GRAY WOODPECKER
Dendropicos goertae
This species is common in African savanna woodland. Its crown is red in adult males and plain gray in females.

EURASIAN WRYNECK
Jynx torquilla
One of two cryptically colored woodpeckers, this species migrates between Europe and Africa, where the Red-throated Wryneck (*J. ruficollis*) lives.

principalis), which may or may not remain in some remote southern North American forests. The last official sighting in the US was recorded in 1944 in the Tensas River region of Louisiana, and in 2021 the Fish and Wildlife Service (FWS) made a proposal to declare the ivory-billed woodpecker extinct. However, experts claim to have collected evidence of the species in Louisiana, thus delaying the FWS's decision. A handful of images of ivory-billed woodpeckers, including one in Audubon's *Birds of America*, reveal the bird's magnificence. Its ivory-white bill was much sought-after, and Indigenous peoples in North America used it for decoration. This resulted in finds of "ivory" bills far to the north of the species' range, in Canada, where animal skins were traded for woodpecker bills at least into the 18th century.

A similar-looking but less renowned species is Mexico's Imperial Woodpecker (*C. imperialis*). If it is still living, it is the world's biggest woodpecker species, but it was last seen in 1956. It was captured at that time on 16mm film, but was otherwise never photographed.

Acorn woodpecker **food stores** are called **granaries**, and may hold as many as **50,000 acorns**

The striking red cap is more prominent on the male woodpecker

Acorns are inserted in specially drilled holes or natural splits in the bark

➤ Winter food store

Acorn Woodpeckers (*Melanerpes formicivorus*) form complex social groups, which create and share stores of acorns in bark or wooden structures, such as log cabins.

TONGUE AND BONE

The hyoid bone is responsible for supporting the tongue, and it sits under the jaw in mammals. In woodpeckers, this bone splits into two, running from the nostrils over the crown of the head and down behind the skull, before curling forward under the throat. It can withdraw the bird's tongue or project it from the bill to probe into holes in bark or anthills. In some species, the tip of the tongue is smooth and sticky; in others, it is armed with tiny barbs.

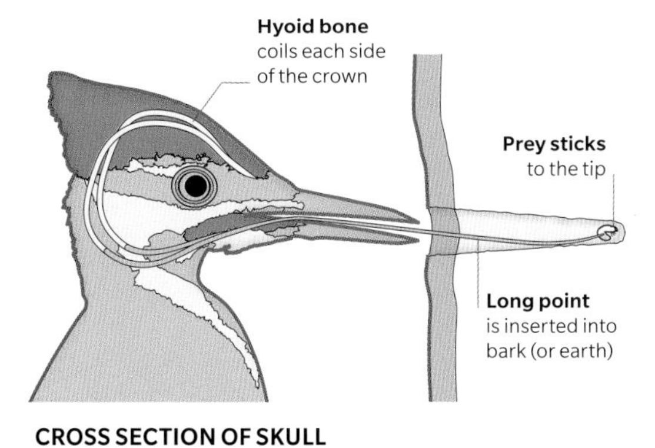

CROSS SECTION OF SKULL

The two-tone bill is black on top and whitish underneath

▲ **Social grouping**
Speckled Mousebirds (*Colius striatus*) form groups of 3–15 birds, with larger flocks of up to 50 gathering to feed.

▲ **Anatomical study**
An early 20th-century plate depicts a White-headed Mousebird (*Colius leucocephalus*) and the heads of a Blue-naped Mousebird (*Urocolius macrourus*) and the pale form of a Speckled Mousebird (*C.s. leucotis*).

Mousebirds

Coliidae

Mousebirds are named for their mouselike appearance and behavior, scurrying along branches and through scrub. Also known as "colies," they are found only in Africa.

The plumage of mousebirds is unusual. Their body feathers are soft, fluffy, and almost hairlike—similar to the fur of a mouse. They also have unusual toes, with strong hooked claws. Their outer toes are reversible, meaning they can point forward or backward. The more usual three toes forward, one back (anisodactyl) is used for perching. Having all four toes pointing forward (pamprodactyl) enables mousebirds to hang upside down from a branch. Having two forward and two back (zygodactyl) helps them climb tree trunks. Mousebirds can run very fast, and with their short rounded wings, they can also fly fast—the Red-faced Mousebird (*Urocolius indicus*) has been recorded flying at speeds of up to 45 mph (70 kph).

In cold weather, mousebirds can become torpid to conserve energy, with their body temperature falling as low as 64.8°F (18.2°C). Research has shown that huddling together in groups of six or more can reduce nighttime energy expenditure by 50 percent. Today, mousebirds are found only in Africa, but fossils reveal that they once lived in Europe and North America.

Order Coliiformes

Families 1

Species 6

Size range 11½–15 in (29–38 cm) long

Distribution Sub-Saharan Africa

Seriemas

Cariamidae

Order	Cariamiformes
Families	1
Species	2
Size range	27½–35½ in (70–90 cm) long
Distribution	South America

Large, long-legged, ground-dwelling birds from central and eastern South America, seriemas stride across dry grasslands or glide through landscapes of dense and uninviting thorny scrub.

At dawn, a far-carrying, yelping duet issues from the Argentine savanna: a pair of Red-legged Seriemas (*Cariama cristata*) are serenading one another. Both seriema species (the other is the Black-legged Seriema, *Chunga burmeisteri*) look as odd as they sound, striding around on absurdly long legs, and their large eyes and sharp, slightly down-curved bill are reminiscent of a raptor.

A closer look at a red-legged seriema reveals two even more unusual features. A tufty crest sprouts from the forehead—the genus name comes from the Tupi language, in which *çariama* means "crested." Above the eyes, what look like eyelashes are not modified hairs as they are with humans, but bristly feathers that protect them from wind-blown dust.

Home birds

In Brazil, there is a saying, "Where there are seriemas, there are no snakes." This refers to the fact that many rural dwellers allow seriemas to live outside their home, sometimes even coming inside to feed on scraps. This attitude is slightly misguided. Seriemas are not particularly avid serpent munchers, but they will kill small prey such as mammals, snakes, lizards, frogs, and birds by beating them on the ground with their bill. In Argentina's Chaco province, black-legged seriemas have been kept as pets, but are also hunted for food and their eggs are harvested.

▼ Odd bird
Seriemas resemble Africa's Secretarybird (*Sagittarius serpentarius*), but they are not closely related. In fact, their closest relatives are falcons.

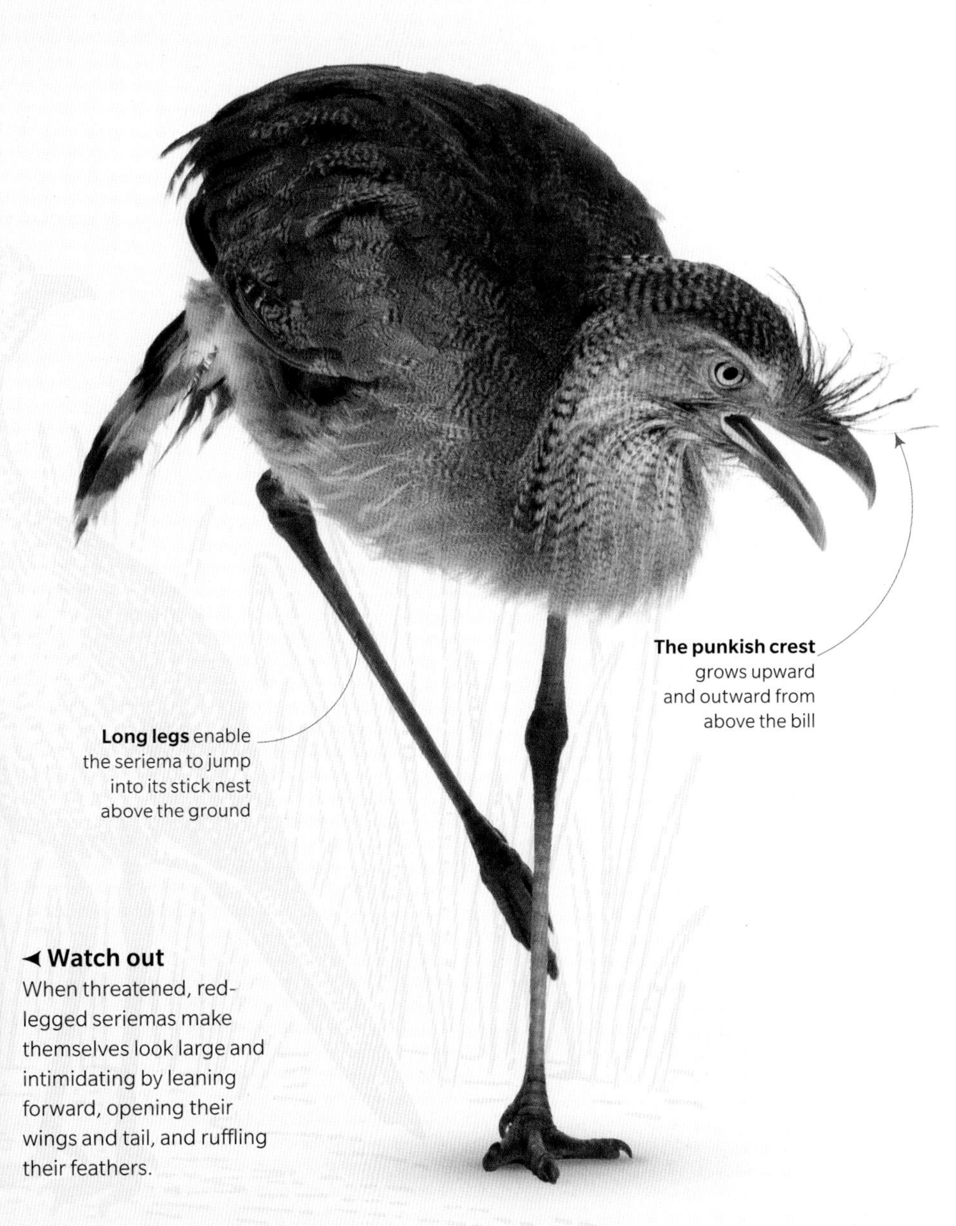

Long legs enable the seriema to jump into its stick nest above the ground

The punkish crest grows upward and outward from above the bill

◄ Watch out
When threatened, red-legged seriemas make themselves look large and intimidating by leaning forward, opening their wings and tail, and ruffling their feathers.

Falcons and Caracaras

Falconidae

This family includes some devastatingly powerful hunters, such as the much-vaunted peregrine falcon, but also some startlingly clever and adaptable generalist foragers in the form of the caracaras.

Pointed wingtips enable fast flight, but mean that large falcons are not adept at soaring

A relatively short and blunt-ended tail completes this bird's aerodynamic outline

Clawed feet will be curled into punching "fists" to make a strike

Nodules known as baffles in the nostrils make breathing easier while diving and have inspired jet-engine design

▼ Locked and loaded
Target in sight, a peregrine falcon prepares to dive on prey below. The force of its high-speed, gravity-assisted strike is often enough to kill its victim.

Falcon-headed god
With the head of a Lanner Falcon (*F. biarmicus*), the Egyptian god Horus carried the Sun in one eye and the Moon in the other, representing power and healing, respectively. This beautiful wall painting from c.1292 BCE adorns the tomb of Pharaoh Horemheb, in the Valley of the Kings.

Less well-known Egyptian falcon-headed gods include
Montu, god of war, and Khonsu, god of the Moon

Hawks and falcons represent an outstanding example of convergent evolution, in that they have independently evolved similar features. The two groups have so much in common—in anatomy and behavior—that it took DNA sequencing studies in the early 2000s to reveal that they are not close relatives. In fact, falcons are much closer cousins to parrots. Their parrotlike traits may not be obvious, but falcons—and especially caracaras—are clever, inquisitive birds, and several falcon species have a long history as close companions of humans.

Strong and intelligent

The large falcons have been prized as falconry birds since the Middle Ages, not only for their power, intelligence, and willingness to work with handlers but also their ability to tackle fair-sized quarry such as grouse and rabbits. Falcons have been important in Bedouin culture for hundreds of years, especially the locally occurring Saker Falcon (*Falco cherrug*), which is the national bird of Oman, Qatar, Saudi Arabia, UAE, and Yemen. More recently, other falcon species from elsewhere in the world have become popular in the Middle East, and the "best" birds (in terms of hunting prowess, rarity, and esthetics) change hands for eye-watering sums of money. In 2021, a White Gyrfalcon (*F. rusticolus*) was sold at auction for a record-breaking 1.75 million Saudi riyals ($466,667).

With their long pointed wings, falcons excel at fast flight and devastating stooping attacks—reaching speeds in excess of 200 mph (320 kph) in the case of the Peregrine Falcon (*F. peregrinus*). Many falcons prey on other birds, which they strike down in midair. Smaller falcons, such as hobbies,

Order Falconiformes

Families 1

Species More than 60

Size range 6–23½ in (15–60 cm) long

Distribution Worldwide except Antarctica

Falcons have a "**tomial tooth**" on their bill, which is used to **disable prey**

are skilled at chasing agile insects, including dragonflies. Forest falcons, mostly found in South America, are broader-winged and resemble hawks from the genus *Accipiter* in their hunting style. Kestrels are a distinct group of falcons found across Africa and Eurasia. They capture mice and other small prey, mainly on the ground, and attack by first hovering to sight their prey, and then dropping like a stone. The Common Kestrel (*F. tinnunculus*) is the titular character of the British film *Kes*, based on the novel *A Kestrel for a Knave* (1968) by Barry Hines. Caracaras are mainly scavengers, but they are also great opportunists and will search on foot for anything edible, sometimes working in groups. In the Falkland Islands, the Striated Caracara (*Phalcoboenus australis*) approaches human visitors for scraps, and loiters around colonies of seals and penguins, ready to attack unattended pups and chicks.

The **Yellow-headed Caracara** (*Milvago chimachima*) picks **parasites** from the fur of **tapirs and capybaras**

◄ Regal pursuit
A 14th-century edition of *The Art of Falconry* depicts the author, Holy Roman Emperor Frederick II, with a trained Saker Falcon (*F. cherrug*) at his side.

AMERICAN KESTREL
Falco sparverius
Despite its kestrel-like hovering habit, this small falcon of the Americas is closely related to the peregrine and other large falcons.

BLACK-THIGHED FALCONET
Microhierax fringillarius
One of the world's smallest birds of prey, this distinctive falconet occurs in Southeast Asia. It is an insect-eater, and sometimes hunts in small groups.

MOUNTAIN CARACARA
Phalcoboenus megalopterus
This striking Andean caracara scavenges on village streets. Like other caracaras (but unlike falcons), it constructs its own stick nest.

One clue to the falcons' taxonomic relationships with other groups is that—unlike hawks—they do not build nests. Instead, like parrots, they nest in tree cavities, on cliff ledges, and sometimes in old nests of other birds. Peregrine falcons will readily take to artificial open nesting boxes, which has helped this globally widespread but often persecuted species to thrive in towns and cities. Urban peregrines often win approval for hunting street pigeons, but their taste for these birds was such a concern during World Wars I and II (when homing pigeons were used as life-saving messengers) that the UK Ministry of Defence ordered a mass cull of wild peregrines.

Today, the Brown Falcon (*F. berigora*) of Australia has gained notoriety for its dangerous behavior. Along with the Black Kite (*Milvus migrans*) and Whistling Kite (*Haliastur sphenurus*), it helps wildfires spread by moving smouldering wood from one location to another. Nicknamed "firehawks," these birds use fire to drive small mammals out of cover, and may also feed on the remains of those that perish in the flames.

SUNSCREEN STRIPES

The dark vertical markings on the cheeks of many falcons, known as malar stripes, might resemble streaks of over-applied makeup. However, they serve a vital purpose when falcons are out hunting in bright skies—they reduce the amount of glare from the sun that is reflected into their eyes. In peregrine falcons, found worldwide in various latitudes, the width of the malar stripe varies according to the average sunlight levels they experience.

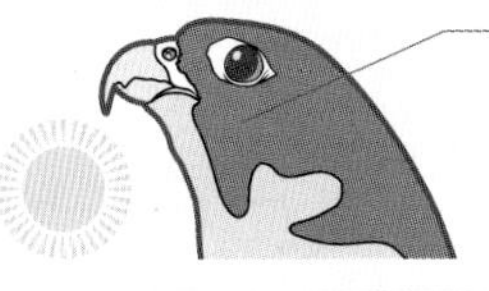

MALAR STRIPE: SUNLIGHT RATIO

" Turning and turning in the widening gyre
The falcon cannot hear the falconer;
Things fall apart; the center cannot hold;
Mere anarchy is loosed upon the world. "

W. B. YEATS, "The Second Coming," 1919

◄ Jack of all trades
Found from southernmost North America to the tip of South America, Crested Caracaras (*Caracara plancus*) are aggressive hunters, scavengers, and thieves —but they also eat fruit.

Parrots and Cockatoos

Strigopidae, Psittacidae, Psittaculidae, Cacatuidae

Instantly recognizable, parrots and cockatoos are among the best-loved birds on Earth. Noisy and sociable, they show an astounding variety, from their size and plumage to their diet and behavior.

With their striking plumage, strong sharply hooked bills, and playful personalities, parrots and cockatoos are among the most popular and easily identifiable birds. Their beauty, intelligence, and ability to mimic human speech has made them highly prized by collectors and as household pets. The largest numbers of parrots live in tropical rainforests and cloud forests, but they are found in a range of habitats.

These birds vary incredibly in size, from tiny parrotlets to majestic macaws. The Psittacidae can boast the largest size range of any bird family: the Manu Parrotlet (*Nannopsittaca dachilleae*) has a length of only 5½ in (14 cm), while the Hyacinth Macaw (*Anodorhynchus hyacinthinus*) can reach 39 in (100 cm). The Psittaculidae, meanwhile, include the world's smallest parrot, Australasia's Buff-faced Pygmy Parrot (*Micropsitta pusio*), which measures just 3 in (8 cm)—smaller than the hyacinth macaw's bill.

Birds of many colors

Parrots and cockatoos display a dazzling array of colors that vary widely even within families. While many members of the cockatoo family (Cacatuidae) are white, some are black or gray, splashed with pink, red, or yellow. Even among the predominantly green New World and African parrots, there

◄ The love bird
Kamadeva, the Hindu god of erotic love, is often depicted riding a parrot. Perhaps due to their longlasting pair bonds, parrots are often associated with love.

► Mutual preening
Blue-and-yellow Macaws (*Ara ararauna*) from South America stay with the same mate for life. Grooming each other not only gets rid of parasites but strengthens the partners' long-lasting bond.

Order Psittaciformes

Families 4

Species More than 390

Size range 3–39 in (8–100 cm) long

Distribution North and South America, Africa, Asia, Australasia, Oceania

are red and blue macaws, while the Golden Parakeet (*Guaruba guarouba*) is startlingly yellow, and the Gray Parrot (*Psittacus erithacus*) has a sophisticatedly silver sheen. Many species are multicolored, including the well-named Rainbow Lorikeet (*Trichoglossus moluccanus*).

Most parrots have a similar body shape: they are chunky, with a big head, short neck, and a large, deep-based, steeply curved bill. They have zygodactyl feet, where two toes face forward and two backward—the latter act as "opposable thumbs" to help the parrot climb and dextrously manipulate its food. The most noticeable difference in body shapes is the shape of the tail. Racket-tails (*Prioniturus* spp.) have a spatula at the tip of their long, bare tail filaments. Many parakeets, lorikeets, and macaws have an extravagantly long tail. In contrast, parrotlets have a markedly short tail.

Parrots primarily eat fruits, nuts, and seeds. Some have different tastes, however. Lories feed on nectar, the brushlike tip of their tongue helping to speed up extraction of the sugar-rich liquid. Pygmy parrots glean fungi and lichen from bark. In New Zealand, the Kākā (*Nestor meridionalis*) supplements its diet with honeydew from scale insects. The world's only flightless parrot, the Kākāpō (*Strigops habroptila*), is strictly vegetarian, but the Kea (*Nestor notabilis*) will feed on meat from carcasses and will even attack live sheep.

Exceptions to the rule

Most parrots form monogamous pairs (macaws even mate for life) and nest in a tree cavity or other hole. But there are exceptions. Monk Parakeets (*Myiopsitta monachus*) nest collectively in massive stick nests built in trees. Burrowing Parrots (*Cyanoliseus patagonus*) excavate elaborate labyrinths in cliff faces. Australia's eclectus parrots (*Eclectus* spp.) breed cooperatively, and females mate with, and are fed by, several males. Among Madagascar's Greater Vasa Parrots (*Coracopsis vasa*), both sexes are promiscuous: the males feed multiple females, and females sing loud complex songs to attract males.

◄ Extinct beauty
Once a common sight across the southeastern US, the last wild Carolina Parakeet (*Conuropsis carolinensis*) was seen in 1910. Its extinction was largely due to human persecution and deforestation.

WHO'S A CLEVER BOY?

Parrots are renowned for their language abilities. A captive gray parrot called Alex was taught by an American academic, Irene Pepperberg, for 30 years. Alex learned to recognize relationships between 100 words, and could identify colors, shapes, and numbers. He could answer questions, count, and even combine established words to make new ones.

▲ Seeking minerals These Red-and-green Macaws (*Ara chloropterus*) visit exposed banks to eat clay. Parrots may eat charcoal, salt, or clay to neutralize toxins in fruit and seeds.

"You be good, I love you. See you tomorrow."

LAST WORDS OF ALEX, THE GRAY PARROT, September 6, 2007

Human fascination with parrots has a long history. There are ancient Indian records of pet parrots from 3,000 years ago. In South America, mummified parrots dating back 900 years have been discovered in the Atacama Desert. Feathers have long been used in ceremonial dress or artifacts, and in Bolivia, villagers still dance wearing an elaborate headdress of macaw feathers shaped into the Sun's rays. In Venezuela, Warao shamans use macaw feathers to decorate a rattle. Elsewhere, people adorn their bodies or clothing with feathers from species as diverse as the Black-capped Lory (*Lorius lory*) in New Guinea and Kuhl's Lorikeet (*Vini kuhlii*) in the Cook Islands. In West Africa, gray parrot feathers are considered an aphrodisiac.

Parrots and cockatoos have long been kept as cage birds, sometimes as a sign of wealth or prestige. Today, about 50 million parrots are kept as pets, and the trade has significantly reduced the wild populations of many species. Paradoxically, then, parrots have sometimes suffered as a result of this enduring fascination.

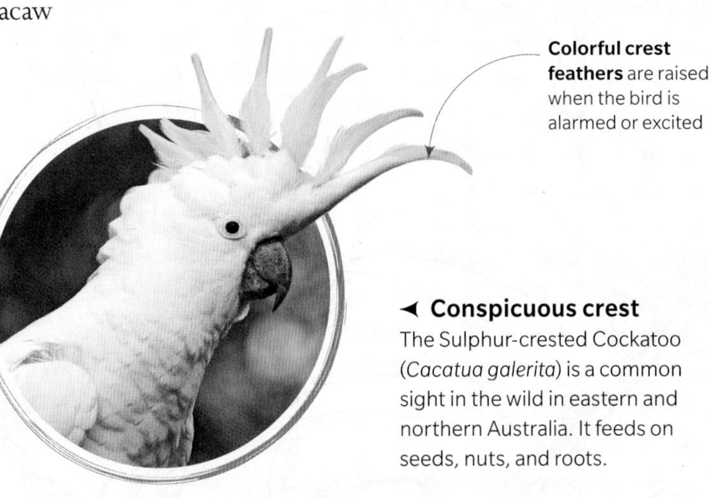

Colorful crest feathers are raised when the bird is alarmed or excited

◄ Conspicuous crest The Sulphur-crested Cockatoo (*Cacatua galerita*) is a common sight in the wild in eastern and northern Australia. It feeds on seeds, nuts, and roots.

ALEXANDRINE PARAKEET
Psittacula eupatria
One of the largest parakeets, this species is from South and Southeastern Asia. Feral birds can be found in Europe and Australia.

KAKAPO
Strigops habroptila
From New Zealand, the nocturnal Kakapo is the world's only flightless parrot. It is also the heaviest, weighing up to 6½ lb (3 kg).

CHAPTER 3

Passerines

More than half of all bird species are passerines, in the order Passeriformes, commonly known as perching birds. All of them have highly specialized slender, four-toed feet. Most of them are also songbirds, thanks to a complex voice box that allows them to create a huge variety of sounds.

Pittas

Pittidae

These delightful forest jewels are very secretive. Much sought-after by birdwatchers, their shy nature means they are usually only seen by the most dedicated.

Noisy pittas use an **"anvil"** of **stone**, or a **branch** or **root,** on which **to break open snail shells**

Pittas have a large head, short wings, long legs, and a short tail. Both sexes tend to be colorful, often with iridescent areas that shimmer like jewels. Pittas are solitary outside the breeding season, but migratory species may form loose flocks. Those that migrate do so at night and can be attracted to lights when migrating. Pittas are found in both temperate or tropical forests, preferring dark areas with plenty of undergrowth. Some species like bamboo thickets, while others choose to be close to water in mangroves or swampy forest. The calls tend to have monosyllabic or disyllabic notes, and some pittas are very vocal.

Despite being elusive, pittas are targeted by the wild-bird trade due to their attractive plumage. The Indian Pitta (*Pitta brachyura*) is often caught by being deliberately attracted to lights. In Vietnam, pittas are also caught for food. In the uplands of Borneo, dried skins of Blue-winged Pittas (*P. moluccensis*) were once used as children's toys.

Order Passeriformes

Families 1

Species More than 40

Size range 5–11½ in (13–29 cm) long

Distribution Africa, Asia, Australia

Yellow and orange streak above the eye

▲ Banded belly
This attractive Malayan Banded Pitta (*Hydrornis irena*) can be identified as a female due to the banded underparts, the male having a blue belly.

Feeding on the ground

Pittas mostly eat invertebrates, such as spiders, grasshoppers, beetles, snails, and termites. They turn over leaf litter and dig into the soil to find food and have been observed with head cocked as if listening for prey. Some species take small frogs, lizards, and snakes. The Mangrove Pitta (*P. megarhyncha*), which has the largest beak out of all the pittas, feeds on crabs.

The males perform different courtship displays, involving bowing, bobbing, and wing-flicking. Both sexes build the domed nest from sticks, grass, leaves, and bark, on or near the ground, often with a front entrance, and both incubate the eggs and care for the young.

► Easy to identify
The black head and underparts, green and blue back, and scarlet undertail make the Rainbow Pitta (*P. iris*) unmistakable. This etching by Elizabeth Gould is from her husband's *Birds of Australia* (1848).

GURNEY'S PITTA
Hydrornis gurneyi
The most endangered pitta, Gurney's Pitta was rediscovered in Thailand in 1986, but has nearly vanished again due to deforestation.

FAIRY PITTA
Pitta nympha
Breeding in Japan, this is the northernmost pitta species and a long-distance migrant, wintering in Borneo. It is rare and its numbers are declining.

NOISY PITTA
Pitta versicolor
The Noisy Pitta is the most common pitta in Australia, and surprisingly given its common name, it is silent outside the breeding season.

Ovenbirds and Woodcreepers

Furnariidae

Order Passeriformes

Families 1

Species More than 310

Size range 10–36 cm (4–14 in)

Distribution Mexico, Central America, South America

The ovenbirds and woodcreepers are remarkable for having evolved an amazing array of characteristics that enable them to inhabit almost every conceivable landscape.

It would be a mistake to dismiss ovenbirds and woodcreepers as dull simply because most species are essentially brown in color. Quite the opposite. Remarkably, this hugely diverse and imaginatively named family occupies every terrestrial habitat in Central and South America: from stony Andean mountains to lush tropical forests, from rocky coasts to scrubby savannas. In these varied landscapes, cinclodes, horneros, and earthcreepers stride along the ground, foliage-gleaners and reedhaunters skulk in the undergrowth, while xenops dangle from slender branches high in the forest canopy.

Fitting the bill

It is their wide range of bill and tail structures that allow ovenbirds and woodcreepers to exploit so many ecological niches. Many species have stiffened tail feathers to support them as they move through vegetation. This is particularly true of

▼ Under construction
This rufous hornero is still building its nest. When completed, the nest will resemble a dome-shaped clay oven (called an "horno" in Spanish), with a slitlike entrance to one side.

The nest is typically built from a mixture of clay, mud, plant material, and dung

Like nearly all ovenbirds, the rufous hornero has plain brown plumage

This hornero is building its nest on a tree limb; other birds may build theirs on a house or lamp post

The intricate **tail feathers** of Des Murs's wiretail more than **double** the bird's **total length**

The threadlike tail is made up of only six feathers; other ovenbird species usually have 12

A pale wingbar is typical of many ovenbirds; it is formed of pale bases to the flight feathers

woodcreepers and treerunners, whose barb-tipped tails help them ascend trees. Ovenbird bills, meanwhile, extend over an impressively wide spectrum—at one extreme, the short, uptilted, chisel-like beak of the White-throated Treerunner (*Pygarrhichas albogularis*); at the other, the lengthy curved bill of the Scimitar-billed Woodcreeper (*Drymornis bridgesii*).

Most ovenbirds and woodcreepers hunt invertebrates by gleaning from leaves or crevices, or rummaging through leaf litter. Larger birds, such as the Ivory-billed Woodcreeper (*Xiphorhynchus flavigaster*), boast a hefty bill capable of catching small vertebrates, such as lizards. The Peruvian Seaside Cinclodes (*Cinclodes taczanowskii*) does things differently, possessing a taste for mollusks and crustaceans gathered from rocky shores as the tide retreats, and even sometimes catches fish.

Feathering the nest

Ovenbirds' nests are similarly varied and remarkable. The Firewood-gatherer (*Anumbius annumbi*) builds a nest that is up to 6½ ft (2 m) long. Several groups of ovenbirds, including leaftossers (*Sclerurus* spp.), excavate burrows in which to conceal their chicks. And the name "ovenbird" itself comes from the football-sized, spherical mud-nests constructed by eight species of horneros (*Furnarius* spp.). The Rufous Hornero (*F. rufus*) often builds its ovenlike nest on walls and fences. Coupled with its fearless nature (birds strut around within feet of people), strong pair bond, and exceedingly loud duets, it is little surprise that this ovenbird has been celebrated as the national bird of Argentina since 1928.

▲ Tail of wonder
Des Murs's Wiretail (*Sylviorthorhynchus desmurii*) lives in scrubby temperate forests on either side of the Andes, in Argentina and Chile.

An assiduous nest-builder, the firewood-gatherer can construct a stick-home ten times longer than itself

SCIMITAR-BILLED WOODCREEPER
Drymornis bridgesii
Named for its strikingly long, decurved beak, this bird differs from other woodcreepers by feeding as readily on the ground as in trees.

PINK-LEGGED GRAVETEIRO
Acrobatornis fonsecai
This Brazilian bird was only discovered in 1994. Its name comes from a Portuguese word for "twigs," which it uses to build its conspicuous nest.

LARKLIKE BRUSHRUNNER
Coryphistera alaudina
Unusual among its family for living almost entirely on the ground, where it sprints around, this bushrunner has a prominent larklike crest on its head.

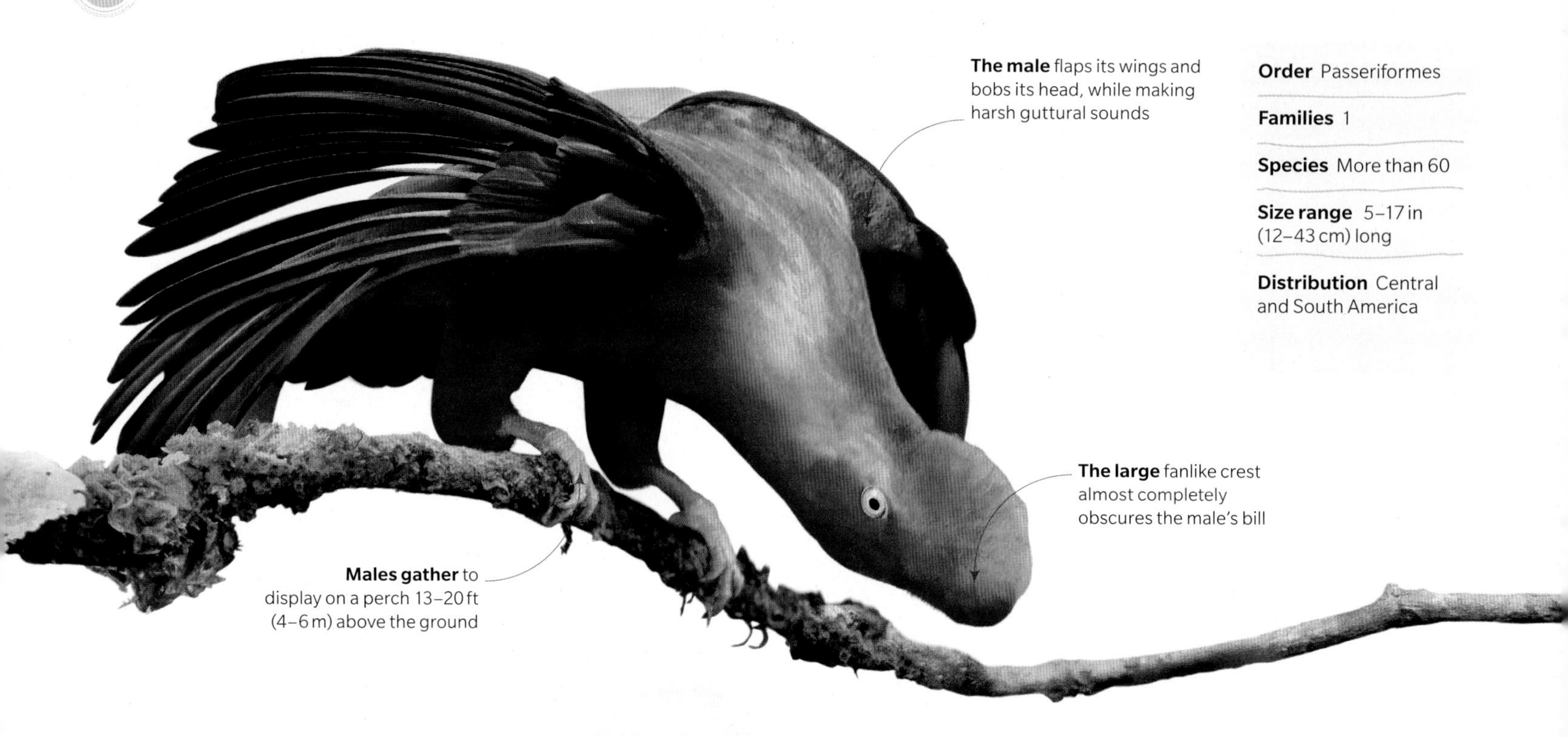

Order Passeriformes

Families 1

Species More than 60

Size range 5–17 in (12–43 cm) long

Distribution Central and South America

Cotingas

Cotingidae

▲ **Bowing to impress**
Andean Cock-of-the-rock (*R. peruvianus*) males jump up and down and bow to attract females.

The cotingas include some of the most dazzling and bizarre birds of the American tropics. Many obtain their brilliant colors from pigments in their fruit diet.

For sheer diversity, glamour, and strangeness, few passerine families match the cotingas. They range from chickadee-sized fruiteaters to crow-sized umbrellabirds. Some species have crests, others wattles, some colorful bare skin, others swallowlike tails. Little unites them except their DNA and the shape of their voice box (syrinx). Males of many species exhibit amazing colors, from red, yellow, and pink to blue and totally white, while the females are typically far less colourful. The Lovely Cotinga (*Cotinga amabilis*) is a prime example of this sexual dimorphism.

Many male cotingas have loud courtship or territorial calls. The White Bellbird (*Procnias albus*) produces the loudest calls of any bird, reaching 125 decibels, while the 116-decibel Screaming Piha (*Lipaugus vociferans*) is the defining soundtrack of the Amazonian forest.

Cotingas are birds of humid lowland and montane forests, and most species largely depend on fruit as their main diet. The Guianan Cock-of-the-rock (*Rupicola rupicola*) has been known to eat 65 species of fruit in a month, while the White-cheeked Cotinga (*Zaratornis stresemanni*) feeds mainly on mistletoe berries.

Many species are polygynous, and males perform in groups called leks, including the cocks-of-the-rock and the Capuchinbird (*Perissocephalus tricolor*), which makes a sound like a malfunctioning chainsaw as it displays. The Dusky Piha (*L. fuscocinereus*) has a remarkable parachute-like flight display below the canopy. Females carry out nesting duties alone.

▲ **Bright and dull**
The brilliant blue and purple of the male lovely cotingas contrast sharply with the drab hues of the female in this illustration from 1857.

➤ All for one
A Blue Manakin (*Chiroxiphia caudata*) flies in to join two other males at a lek. Their group display includes each bird taking turns to jump up and flutter over its dance partners. Only the alpha male mates with the female.

Manakins

Pipridae

Order	Passeriformes
Families	1
Species	More than 50
Size range	3–6 in (7–15 cm)
Distribution	Central and South America

Confined to the American tropics, from southern Mexico to northern Argentina, this family of small forest understory birds is most famous for its elaborate dancing displays.

Manakins are small and rotund, with a large head, short bill, and broad, rounded wings. They flit among branches to pluck fruit, their main food, and catch the occasional insect on the wing. Like cotingas, manakins show extreme sexual dimorphism: females have dull, inconspicuous plumage and males are boldly marked, often with bright colors, particularly red and blue.

Roles and rituals

Female manakins have sole responsibility for nesting duties, such as building simple cup-shaped nests and feeding their young. Since the males contribute nothing beyond the mating process, females are highly selective of their mates. This has led to extreme competition for their attention, especially in display, and only a small proportion of males are chosen to be a reproductive partner. In the breeding season, males may spend 90 percent of their time in leks, and many will attend one almost every day of the year. Members of a lek either gather closely together or place themselves within hearing distance of each other in what is known as an exploded lek. Some species use a branch to clear a patch of ground to use for these displays, while others perform on moss-carpeted logs after removing any leaves or other objects from their surfaces.

Manakin "dances" include somersaults, bowing, sidling along perches, fluttering, jumping, hovering, and ruffling colorful feathers. The Red-capped Manakin's (*Ceratopipra mentalis*) backward sidling has been compared to the late performer Michael Jackson's famous "moonwalk" maneuver. Many manakins make loud mechanical wing-claps. Female Club-winged Manakins (*Machaeropterus deliciosus*) are an especially tough audience—they listen so carefully that they assess the resonant quality of the males' claps.

The bold pattern of the wire-tailed manakin allows competing males to see each other at an understory lek

➤ Taking it on the chin
The male Wire-tailed Manakin (*Pipra filicauda*) will court the female by tickling her chin with these wiry tail filaments.

Antpittas and Antthrushes

Grallariidae, Formicariidae

Order Passeriformes

Families 2

Species More than 80

Size range
4–11 in (10–28 cm) long

Distribution
Mexico, Central and South America

Even today, scientists know very little about these secretive and beguiling birds, which spend their lives hidden within the dense forests of Central and South America.

◄ Living the high life
The Tawny Antpitta (*Grallaria quitensis*) lives in open country as high as 15,000 ft (4,500 m), as shown in this illustration from 1849.

With plump, rounded bodies and short tails, antpittas have been described as looking like "tennis balls with legs." Combined with large eyes, short wings, and an elaborate hopping motion, these birds have a cartoonish appearance that makes them instantly recognizable.

These characteristics make antpittas highly sought-after targets for birdwatchers. But getting a view of an antpitta—or one of the equally elusive cocked-tailed antthrushes—is far from easy. These birds are usually much more readily heard whistling loudly than they are seen.

Close encounters

Some enterprising locals have created thriving businesses out of habituating birds through hand-feeding. Originally practiced at a reserve in northern Ecuador, the technique has spread widely, and species once considered impossible to spot can now be watched eating at point-blank range. With tourists willing to pay good money for the experience, it's a win-win situation—conservation-minded locals earn funds to protect the habitat, and birdwatchers get to see their prize.

Large eyes ensure good vision in the low light conditions of the forest floor

Long, powerful legs are an antpitta's primary mode of transport—they prefer to hop rather than fly

Despite their popularity, most aspects of antpitta and antthrush biology remain mysterious, and they are among the least-known families of birds in the Americas. For example, nests have been found for less than half the antpitta species.

◄ Shy explorer
The Chestnut-crowned Antpitta (*Grallaria ruficapilla*) has long legs, which it uses to explore the the forest floor. It hides in dense vegetation, rarely venturing out into the open.

GIANT ANTPITTA
Grallaria gigantea
Restricted to the cloud forests of the Andean foothills in Ecuador and Colombia, this is the largest of the antpittas.

OCHRE-BREASTED ANTPITTA
Grallaricula flavirostris
This tiny antpitta is known for its "dance," in which it swings its body while holding its head completely still.

RUFOUS-BREASTED ANTTHRUSH
Formicarius rufipectus
Found in cloud forests from Costa Rica to Peru, this stocky antthrush has the characteristic short, cocked tail of its family.

Antpittas use their strong bill to toss leaf litter aside to find earthworms, ants, and other insects

NEW DISCOVERIES

New species of antpittas and antthrushes continue to be found. These include the striking Jocotoco Antpitta (*Grallaria ridgelyi*), which was only discovered in 1997 and has a tiny range in Ecuador and northeast Peru. A reserve has been specially created to preserve the bird and its habitat.

JOCOTOCO ANTPITTA

Order Passeriformes

Families 1

Species More than 440

Size range 2½–15¾ in (6.5–40 cm) long

Distribution North America, South America

◄ Quick to anger
The great kiskadee is an aggressive and bold tyrant. Here, one is flying up to attack a Clay-colored Thrush (*Turdus grayi)* that has ventured into its personal space.

▲ Colorful tyrant
This illustration from *Birds of La Plata* (1920) depicts an outstanding exception to familial drabness. The many-colored Rush Tyrant (*Tachuris rubrigastra*) is known in some parts of South America as *siete colores* (meaning "seven colors" in Spanish).

Tyrant Flycatchers

Tyrannidae

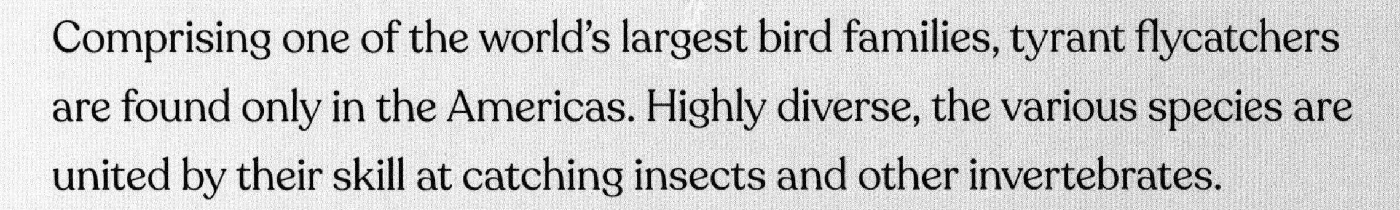

Comprising one of the world's largest bird families, tyrant flycatchers are found only in the Americas. Highly diverse, the various species are united by their skill at catching insects and other invertebrates.

The tyrant flycatchers is one of the dominant bird families of the Americas, comprising, for example, almost a fifth of all the landbird species of South America. Many are easily overlooked forest-dwellers, with dozens of forms hidden away in the canopy. A significant number have unremarkable plumage and they generally lack songs that are pleasing to human ears.

Among the multitudes of birds in this family, it is difficult to define exactly what a tyrant flycatcher is, and the composition of the family frequently changes as DNA studies shuffle the pack. However, the members are essentially insect-eaters and obtain their food by sallying. Some leave a perch to chase an aerial insect and then return to a perch—a technique known as hawking. Others leave their perch but snatch food on the ground, while other species sally toward an insect on a leaf or branch, which they then catch in the bill. This latter method is known as sally-gleaning.

Most species in the family have a bill that is broadened at the base, and often flattened, which gives the mandibles a wide surface area to close upon the prey. Most species also have bristles at the base of the bill, which may protect the face or may have a tactile function.

A flycatcher for most habitats

Tyrant flycatchers occupy almost every habitat in the Americas, from the Arctic Circle to the interior of humid equatorial forests. In some parts of the lowlands of the Amazon Basin, 80 species from this one family can be found at a single location, each occupying its own microhabitat. They variously inhabit the highest canopy to mid-levels, with some preferring vines or other specific types of vegetation, and from bright light to deep shade. Elsewhere, there are also species for nearly every habitat—ground tyrants (*Muscisaxicola* spp.) forage on the ground in treeless terrain, while the Torrent Tyrannulet (*Serpophaga cinerea*) perches on rocks in fast-flowing rivers, the Cliff Flycatcher (*Hirundinea ferruginea*) prefers rocky outcrops, and the Sierran Elaenia (*Elaenia pallatangae*) lives in high-altitude montane forest in northwestern South America. And different species are found at different altitudes in the Andes.

Famous tyrants

Despite the general obscurity of tyrant flycatchers, some species are very well known and others are highly colorful and conspicuous. In North America, the much-loved Eastern Phoebe (*Sayornis phoebe*) is a popular sign of spring and often nests on bridges, while the spectacular Scissor-tailed Flycatcher (*Tyrannus forficatus*) is the state bird of Oklahoma. The US statesman Benjamin Franklin was a great admirer of the

▼ Agile flier
The scissor-tailed flycatcher obtains most of its food during aerial forays to snap up insects. It migrates from the southern US to Central America.

Marsh dwellers
The White-headed Marsh Tyrant (*Arundinicola leucocephala*) is unusual among the tyrant flycatchers in that the male (above) and female (below) look entirely different.

aggressive Eastern Kingbird (*T. tyrannus*), which attacks anything from hawks to humans (or even in at least one confirmed case, a light airplane) to defend its territory. The Great Kiskadee (*Pitangus sulphuratus*) is abundant throughout tropical America, even in big cities such as São Paulo, Brazil. Other conspicuous species in this family include the widespread, gaudy Vermilion Flycatcher (*Pyrocephalus obscurus*), the brilliantly white monjitas (*Xolmis* spp.), and the Streamer-tailed Tyrant (*Gubernetes yetapa*).

Limited repertoire

Tyrants are not famous songsters; they have a simple sound-producing syrinx and do not learn their songs (they are innate), which limits the variety. Many, though, have a decent vocabulary, as much as 12 sounds in the case of the Great Crested Flycatcher (*Myiarchus crinitus*). Almost all have a specific dawn song uttered only in twilight.

Most tyrant flycatchers are territorial and form monogamous pair bonds. They build a great variety of nests, from simple cups to elaborate spherical structures with side entrances. Some conceal their nest with moss or vines and some build close to stinging insect colonies for protection.

Before **setting off on** migration, **scissor-tailed flycatchers** roost in flocks of up to **1,000 birds**

GREAT CRESTED FLYCATCHER
Myiarchus crinitus
The secretive Great Crested Flycatcher feeds high up in the canopy foliage of mature forests, catching most of its food by sally-gleaning. It breeds in eastern North America.

STREAKED FLYCATCHER
Myiodynastes maculatus
Widespread over much of South America, the Streaked Flycatcher uses its heavy bill to catch large insects such as cicadas, beetles, and wasps. It also consumes berries.

CURIOUS LINING

Several members of the genus *Myiarchus*, including the great crested flycatcher, make bulky cup nests in cavities and are famous for using an unusual material in the construction —sloughed off snake skins. They collect the skins and weave them into both the main cup and the nest lining. This behavior is not universal, and there is some evidence it occurs in areas where there is a high level of predation from flying squirrels.

EGGS IN SKIN-LINED NEST

➤ Brought to life
This 19th-century painting of a pair of superb lyrebirds by the naturalist John Gould is based on a taxidermist's specimen that was exhibited the British Museum.

Order Passeriformes

Families 1

Species 2

Size range 31½–39 in (80–98 cm) long

Distribution
Eastern Australia

Shy singer
Albert's lyrebird has a shrinking range in montane forests above Australia's Gold Coast. It is seldom seen, but males are very vocal during the breeding season.

Albert's lyrebird is named after the husband of the British monarch Queen Victoria

Lyrebirds

Menuridae

Australia's lyrebirds have some of the most advanced and complex vocal displays of all birds, mixing a variety of mimicked sounds into a repertoire of unique songs.

Lyrebirds are such good **mimics** that they are able to **fool other birds**

An Indigenous Australian legend recounts that the lyrebird was peacemaker in a dispute between all animals. The spirits gave it a vocal repertoire enabling it to communicate with all creatures, while the frog, the cause of the trouble, was reduced to making a simple croak.

Both male and female lyrebirds are renowned for their expert mimicry of birds and mammals, and can even mimic mechanical sounds with uncanny accuracy. The male Superb Lyrebird (*Menura novaehollandiae*) has an especially complex courtship routine of song and dance in a cleared arena on the forest floor, quivering his lacy tail plumes spread high over his head. Vocal mimicry increases with age, but birds in fragmented forest habitats have a smaller repertoire. It has been found that a male creates the sound of a mixed flock of birds mobbing a predator, as if to suggest to a female that he will offer protection. This is not the "honest" declaration of fitness usually associated with bird song, but a form of deception. Its exact benefits are still unclear.

Of the two species, the superb lyrebird is still common in parts of its southeastern Australia range where suitable forest remains, while Albert's Lyrebird (*M. alberti*) is confined to a small area of forest near the Gold Coast, where it is protected.

◄ **Emblematic bird**
The superb lyrebird is featured on emblems and coins in Australia, especially for the states of New South Wales and Victoria.

Bowerbirds

Ptilonorhynchidae

Bowerbirds take their name from the intricate display structures called bowers that males of many species build and decorate profusely with colored items to attract mates.

Order Passeriformes

Families 1

Species More than 20

Size range 22–40 cm ($8\frac{1}{2}$–16 in) long

Distribution Australasia

The feathers that partially cover the base of this bird's bill inform its family name Ptilonorhynchidae ("feather-bill family")

Adult males have a distinctive violet-blue iris

Confined to Australia and New Guinea, bowerbirds are plump, stocky birds with rounded wings. Some are plain-colored, while others have bright yellow, orange, or glossy black plumage. Most are not musical, making harsh clicking, buzzing, and chirping sounds, but the Satin Bowerbird (*Ptilonorhynchus violaceus*) whistles a song. Many are accomplished mimics of birds, human voices, and machinery. They feed on fruit as well as flowers, buds, and leaves, and eat a wide range of insects during the breeding season.

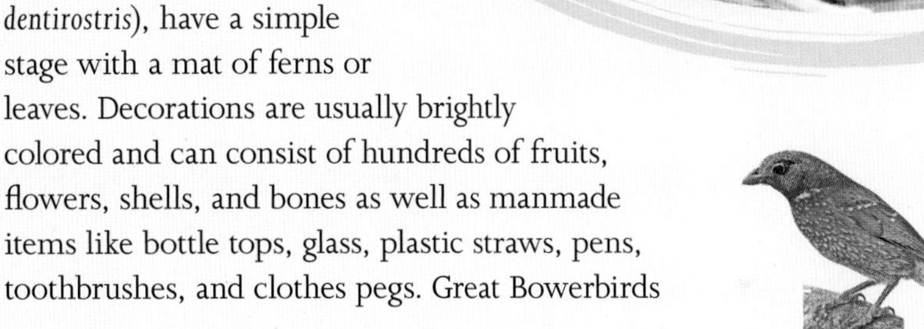

Avenue bower
This 1898 illustration shows a female Great Bowerbird (*Chlamydera nuchalis*) inspecting a male's bower that is decorated with white stones, bones, and shells. To attract her further, the male displays a pink crest, which is usually hidden on his nape.

Decorative displays

In bower-building species, males are polygynous, mating with as many females as they are able to attract to their display arena. There are three different types of bowers: avenue, maypole, and display court.

Avenue bowers are made by the satin bowerbird, the four *Sericulus* species, and the five *Chlamydera* species, and consist of a simple avenue with a wall of thin sticks on each side. Maypole bowers are built by the five *Amblyornis* species and the Golden Bowerbird (*Prionodura newtoniana*) around one or two sapling poles. Some larger bowers resemble a hut, and one huge double maypole bower was 6½–10 ft (2–3 m) tall, 10 ft (3 m) wide, and was used for 40 years. Display court bowers, such as those built by the Archbold's Bowerbird (*Archboldia papuensis*) and Tooth-billed Bowerbird (*Scenopoeetes dentirostris*), have a simple stage with a mat of ferns or leaves. Decorations are usually brightly colored and can consist of hundreds of fruits, flowers, shells, and bones as well as manmade items like bottle tops, glass, plastic straws, pens, toothbrushes, and clothes pegs. Great Bowerbirds (*Chlamydera nuchalis*) arrange their objects by size, with the largest objects farthest away to create an illusion, making the male appear larger. Courtship displays can involve wing drooping or flicking, tail fanning, hiding, strutting, and posturing. In Australia, bowers are often a source of civic pride and many are tourist attractions.

> " These highly decorated halls of assembly must be regarded as the most wonderful instances of bird architecture yet discovered. "
>
> JOHN GOULD, *Handbook to the Birds of Australia*, 1865

Bower-builders' nests are made by the female, who also incubates the one to three eggs and takes care of young. Nests are usually a base of sticks, with an open cup lined with softer leaves or ferns. Male catbirds (*Ailuroedus* spp.) do not build bowers, and both sexes help build their deep-cup nests.

◄ Building attraction
The male satin bowerbird constructs his bower by interweaving sticks. He decorates it with objects, preferring bright blue colors but using yellow and white, too. He will steal some of these items from other bowers.

GREEN CATBIRD
Ailuroedus crassirostris
Named for their catlike wailing call, catbirds are monogamous and mate for life. Green Catbird males and females look the same and are found in subtropical forests on Australia's east coast.

TOOTH-BILLED BOWERBIRD
Scenopoeetes dentirostris
The distinctive toothed bill of this bird is used for cutting and moving leaves: up to 100 are cut and displayed as a stage mat, with their pale undersides facing up.

Exception to the rule
Wallace's Fairywren (*Sipodotus wallacii*) lacks the vivid coloration of other family members, as shown by John Gould in c.1875.

With spangles of blue on a black crown, males and females are similar in appearance

Fresh tail feathers are vivid blue, but fade to brown before being replaced in the annual molt

Fairywrens

Maluridae

Widespread and familiar across Australia, fairywrens occupy a range of habitats, including grasslands, woodlands, and coastal scrubland. A few species are also found in New Guinea.

Fairywrens are tiny birds, commonly associated with the strikingly colorful plumage of the males of some species. The vivid hues include bright blue, purple, and red, but this conspicuous coloring is short-lived, as each year the males molt into a dull brown or "eclipse" plumage outside of the breeding season. Confusingly, "wren parties" of mostly brownish birds can be found in the summer, comprising a territorial breeding pair and a number of immature males, who help rear the first brood of chicks. This first brood, by six weeks old, then take their turn to feed later chicks. In this way, a pair may rear several broods in short order.

Research reveals that males sing to their eggs, and their chicks hatch already recognizing their individual song. Females also call to their eggs using a unique series of sounds, and chicks are able to learn this special phrase and incorporate it into their food-begging calls, thereby reinforcing family bonds. These connections become very important when fairywrens are threatened by predators or in some kind of distress. Nearby birds are more likely to risk helping those that give recognized family calls—by trying to distract the predator, for example—rather than those that are not part of their particular fairywren community.

➤ **Brilliant blues**
Intense black and dazzling blue plumage makes the male Superb Fairywren (*M. cyaneus*) one of Australia's favorite birds.

Order Passeriformes

Families 1

Species More than 30

Size range 4½–6½ in (12–16 cm)

Distribution Australasia

Most species of fairywren live in low scrubby vegetation, but they are also found in suburban areas, where they have been known to "fight" their reflections in windows. The Lovely Fairywren (*Malurus amabilis*) is the most arboreal family member, and prefers shrubby thickets at forest fringes. Purple-crowned Fairywrens (*M. coronatus*) remain with one mate for most of their lives, while other species, such as the Splendid Fairywren (*M. splendens*), form a single pair bond but also mate with other individuals. To attract a mate, male splendid fairywrens sing when predators are close by, as an act of bravado.

VARIATIONS ON A THEME

Loud and rapid, fairywrens' songs comprise seemingly simple phrases based on quick, tinny, trilling notes. Depending on the species, the sound can be harder or softer, or more or less metallic, which makes identification a challenge. Unusually, females sing just like the males. In song, their raised tail shimmers, and their head moves from side to side, adding an apparent variation in volume and pitch.

FEMALE SUPERB FAIRYWREN

" Sometimes [a male] will add to his glamour by carrying a flower in his beak. "

SIR DAVID ATTENBOROUGH, *The Life of Birds*, 1998

The small bill is ideal for foraging on the ground for insects, seeds, and fruits

The round body shape and slender cocked tail recall the unrelated wrens of Europe and America

PURPLE-CROWNED FAIRYWREN
Malurus coronatus

The largest of the *Malurus* species, this fairywren is found in areas of tropical northern Australia in riverside woods, thickets, and marshland.

RED-BACKED FAIRYWREN
Malurus melanocephalus

The smallest of the *Malurus* species, this fairywren is uniquely chocolate-brown to black, with a fiery red back and a relatively short perky tail.

▲ Flight and fight
New Holland Honeyeaters (*Phylidonyris novaehollandiae*) squabble over everything from food to mating. Areas with good nectar sources often resemble a battlefield.

TALKING TUI

The largest honeyeater species in New Zealand, the Tui (*Prosthemadera novaeseelandiae*), has an exceptional voice, and both sexes sing a highly varied song with much mimicry. Captive tuis were kept by Māori and taught to imitate both compliments and insults.

WILD ADULT

Order Passeriformes

Families 1

Species More than 190

Size range 3½–20 in (9–50 cm) long

Distribution Australasia

Nectar lovers
A trio of honeyeaters are depicted in this 1890–91 illustration: Pied (*Certhionyx variegatus*, top), Regent (*Anthochaera phrygia*, right), and Yellow-faced (*Caligavis chrysops*).

There are almost 80 species of honeyeaters in Australia, where they are key pollinators

Honeyeaters

Meliphagidae

One of the three large nectar-consuming bird families in the world, along with the hummingbirds and the sunbirds, honeyeaters play an important role fertilizing many species of flowering plants.

Honeyeaters have a unique tongue with a brushlike tip that absorbs nectar by capillary action. This allows them to sponge up nectar from a wide surface area, and also to lap up insect secretions such as honeydew. The bill is often decurved, and the tongue protrudes beyond the bill tip—by as much as ¾ in (2 cm) in the Red Wattlebird (*Anthochaera carunculata*). The base of the tongue is curled to form a tube, and honeyeaters can lap at a rate of 6–12 times a second. All species eat at least some insects; a few also eat fruit.

Honeyeaters have spread into every habitat in Australasia, from deep desert to rainforest. They are slender-bodied, with long wings, and their strong legs are ideal for gripping branches. Noisy and confrontational, honeyeaters frequently fight over nectar sources. However, a few species exhibit a specific nonaggressive behavior in which they gather in groups of 2–40 individuals, making bowing movements and repeated vocalizations. Their actions are reminiscent of Indigenous Australian ceremonial dances known as "corroborees," and the birds' congregation has been given the same name.

Most honeyeaters hold breeding territories and are monogamous, but the four miner (*Manorina*) species have complex social arrangements. Both sexes can be polygamous, and many individuals may help out at each nest.

CARDINAL MYZOMELA
Myzomela cardinalis
This species lives on island groups in the southwest Pacific. Males have a crimson head and black body, while females are duller, with grayish and reddish-brown coloration.

NOISY FRIARBIRD
Philemon corniculatus
Many honeyeaters have patches of bare skin on their head. The friarbirds' bald patch and drab plumage gave rise to their common name.

Whistlers, Bushshrikes, and Cuckooshrikes

Pachycephalidae, Malaconotidae, Campephagidae

These three distinct families of songbirds from Africa, Asia, and Australasia are united by their loud voices and a powerful, hook-tipped bill that is used to dismember tough invertebrate prey.

Order Passeriformes

Families 3

Species More than 200

Size range 5–15 in (12.5–38 cm) long

Distribution Africa, Asia, Australasia

Most whistlers are found on islands across the Indo-Pacific, with the greatest diversity in New Guinea, but some species are also found in Australia—notably, the Gray Shrikethrush (*Colluricincla harmonica*), which is one of the country's finest songbirds. Whistlers are often dominant voices in the dawn chorus, and birds have even been known to sing when sitting on the nest. Their loud, whistling songs can be heard as far as 1,640 ft (500 m) away.

Without their loud voices, whistlers could easily be overlooked. They spend most of their time in the forest canopy gleaning for insects, often in a deliberate, unhurried manner. The Australian Golden Whistler (*Pachycephala pectoralis*) will typically remain on a perch and methodically peer upward to check leaves for insects. Some species perform a hunting method called sally-striking, where they launch themselves from a branch when prey, such as nonflying insects or spiders, comes within striking range. Whistlers have a broad head, strong legs, rounded wings, and a powerful bill, which is ideal for dealing with the harder exoskeletons of larger insects. Some species, notably shrikethrushes, also eat small mammals, frogs, and lizards. And the White-bellied Whistler (*P. leucogastra*) forages for small crabs and mollusks in the mangroves.

Singers with a hook

Bushshrikes are almost entirely confined to Africa and include some of the continent's most colorful birds. Several species, such as the aptly named Many-colored Bushshrike (*Chlorophoneus multicolor*), also exhibit different color morphs. They have a thick, notched bill with a hooked tip—like that of shrikes (family Laniidae)—which is used to glean invertebrates unobtrusively in the canopy and also in thick vegetation at any level. Unlike shrikes, however, bushshrikes do not swoop down on prey from a lookout perch.

Like whistlers, bushshrikes are also renowned for their songs, which can be loud. In southern Africa, for example, the Gray-headed Bushshrike

▼ Thunderbird
The Australian golden whistler produces melodic whistles, especially at dawn, and is sometimes prompted to sing by loud bangs, including thunderclaps.

◄ Bright and beautiful
Scarlet Minivets (*Pericrocotus speciosus*) are highly sociable birds from southern Asia. They have red, orange, or yellow feathers, depending on their location. In this pair, the male is scarlet and the female is golden-yellow and gray.

When foraging, the bird perches on a branch to search for insects

The family name Pachycephalidae effectively means "thick head."

In 2023, **a species of whistler** in New Guinea **was discovered to be poisonous**, largely due to **toxins in the beetles** it eats

The Bengali name for minivets translates to "seven girl friends" and is due to their sociable nature

(*Malaconotus blanchoti*) is called "Spookvoël," because its song is "ghostly" and haunting. Several species are renowned for their duets, in which males and females of a pair sing a phrase so close together they sound like a single bird. Male and female Slate-colored Boubous (*Laniarius funebris*) initially learn their respective parts from a same-sex tutor. They combine these into a unique, pair-specific song that remains the same for life. Duetting is thought to be an important component in keeping the pair together, even when both birds are hidden in dense vegetation. Bushshrikes are territorial and monogamous, and they build compact cup-shaped nests within dense vegetation in which remarkably small clutches of often just two or three eggs are laid.

Hairy diet

Cuckooshrikes have, like whistlers, found their way to many Indo-Pacific islands from Africa, Asia, and Australia. They are so-called for the resemblance of some species to small cuckoos (family Cuculidae), and for their widespread habit of consuming large numbers of caterpillars, including hairy ones.

Many cuckooshrikes are gray or white, but the minivets (*Pericrocotus* spp.) of Asia are startlingly different, with bold plumage, often with intense red or yellow. These slim, long-tailed birds are familiar to many people in tropical Asia for their noisy flocks. It is typical for many species, even those that live in pairs, to join mixed parties of other forest birds for protection against predators when foraging.

➤ Matched pair
The male and female Yellow-crowned Gonolek (*Laniarius barbarus*) look identical and once paired remain together. They keep in close contact with remarkable song duets that sound like a single bird.

BOKMAKIERIE
Telophorus zeylonus
This southern African bushshrike species is named after its call. The male and female sing "bok-bok" in duet, and songs often vary from place to place, showing local dialects.

CRIMSON-BREASTED SHRIKE
Laniarius atrococcineus
In common with many members of the bushshrike family, this shrike is a skulker, keeping to dense vegetation—in this case, in the thorn scrub of southern Africa.

RUFOUS WHISTLER
Pachycephala rufiventris
This species is unusual among its family in showing migratory behavior. Eastern Australian populations fly south in the spring and head north again in winter.

▲ **Fruit eater**
Cuckooshrikes are known for devouring large insects. However, this Stout-billed Cuckooshrike (*Coracina caeruleogrisea*) from New Guinea, illustrated by John Gould in 1876, also eats fruit, which it usually obtains in the forest canopy.

Cuckooshrikes, much like whistlers and bushshrikes, build intricate cup-nests, often in forks in trees and bushes. The nestlings are often barred or spotted, providing camouflage that allows them to blend into the surroundings with amazing effectiveness. Adults have longer wings and a smaller head than whistlers or bushshrikes, but share the same powerful hooked bill and diet of large, hard-bodied insects. They also have bristles at the base of the bill to protect the face from damage by struggling prey.

An oddity of the family is that these birds have stiffened, spinelike feathers on the lower back and rump that can be erected. It is possible that these feathers are shed as a form of defense if the bird is attacked. Cuckooshrikes also have powder-downs, which are feathers that disintegrate into a powdery substance that they mix with the preen oil they produce from a gland at the base of their tail, and use this mixture to help maintain their plumage.

Shrikes

Laniidae

Shrikes are territorial and carnivorous by nature, and have a robust beak with a hooked tip (like birds of prey) that helps them dismember insects and small reptiles, birds, and mammals.

Order Passeriformes

Families 1

Species More than 30

Size range 6–20 in (16–50 cm) long

Distribution North America, Europe, Asia, Africa

▲ Sherborne Missal Created in c.1400 for an abbey in Dorset, this manuscript depicts English birds, but this shrike appears to be from Spain.

Shrikes are remarkably predatory for their size. They hunt large insects and small rodents, as well as reptiles and birds, moving stealthily or watching from an exposed perch. A shrike the size of a thrush, perhaps 7 in (18 cm) long, can take a warbler half as big. Their target is killed quickly, and then often taken to a thorn bush or spiky fence and impaled. Although armed with sharp claws, a shrike's feet lack the strength of a bird of prey, and impaling its catch helps the bird tear the food into smaller pieces. Shrikes are well known for the way in which they store their prey. It may be kept for future meals, and studies have shown that storing food in a "larder" is especially useful when there are chicks to feed and the weather is poor.

Eye-catching but elusive

The majority of shrikes fall into two groups. Larger species, such as Eurasia and Africa's Great Gray Shrike (*Lanius excubitor*) and the very similar Northern Shrike (*L. borealis*) of North America, often have striking black, white, and gray plumage. They frequently perch on wires

" The Shrike . . . possesses the most undaunted courage, and will attack birds much larger than itself, such as the Crow . . . "

THOMAS BEWICK, *A History of British Birds*, 1797

◄ Social gathering African Northern White-crowned Shrikes (*Eurocephalus ruppelli*) sometimes gather in small groups, looking to follow ground hornbills to snatch insects they disturb.

Impaling the lizard helps the shrike dismember it for eating

or high branches, yet they also spend extended periods on the edge of a bush and become remarkably unobtrusive. Smaller species, such as the Red-backed Shrike (*L. collurio*), often prefer thickets and rough gorse. They can be quite difficult to find, but perseverance has its rewards, because they, too, are beautifully patterned—ancient Egyptian wall paintings that adorn the tomb of Khnumhotep II at Beni Hasan depict a Red-backed Shrike and a Masked Shrike (*L. nubicus*) with their distinctive eyestripes. Other species, such as Africa's Yellow-billed Shrike (*L. corvinus*) and Magpie Shrike (*L. melanoleucus*), are known for their exceptionally long, graduated tails.

Shrikes form pair bonds and defend their territory from other pairs. Females are often attracted by a well-stocked larder, and males present food to the female as part of their courtship ritual. They then raise their offspring together. In Africa's white-crowned shrikes (*Eurocephalus* spp), fledglings from a first brood may help their parents feed a second one.

➤ Butcher bird
This male red-backed shrike has found the perfect branch of thorns on which to store his butterfly and lizard "kebab" for later consumption.

Like falcons, shrikes have a "tooth" on the cutting edge of their bill to help kill prey

GRAY-BACKED SHRIKE
Lanius tephronotus
This largely insectivorous bird occupies high-altitude forest clearings in South and Southeast Asia, and moves lower into cultivated areas in winter.

LOGGERHEAD SHRIKE
Lanius ludovicianus
Predatory on small birds, rodents, reptiles, and insects, this mainly North American shrike sings a quiet, rhythmic song.

Fantails

Rhipiduridae

These small birds are delightful to watch, flitting through the foliage and fanning and waving their long tails, which flash white as they forage for insects.

Order Passeriformes

Families 1

Species More than 60

Size range 4½–8 in (11.5–21 cm)

Distribution Asia, Australasia, Oceania

⮟ Feathers fanned
The Rufous Fantail (*Rhipidura rufifrons*) feeds in forests and woodland, rarely perching as it constantly moves among the branches in search of insects.

While defending their nestlings, willie wagtails have been known to attack humans, cats, dogs, eagles, turtles, and tiger snakes

Fantails are usually seen singly or in pairs and are known to be inquisitive and tame. Rarely keeping still even when perched, these birds will wag their tail from side to side, spin around, or rock to and fro. Mainly blue, black, rufous, and white, most fantail species are slim, long-tailed, and similar-looking across both sexes, except in the Black Fantail (*Rhipidura atra*) of New Guinea, where the male is all black and the female rufous.

Gossip and ghosts

Fantails are usually quiet, but have simple songs and squeaky calls. Some people believe that hearing one sing brings good luck, while in Indigenous Australian folklore, the Willie Wagtail (*R. leucophrys*) is regarded as a bringer of both good and bad news, a gossip, a liar, and a stealer of secrets. In parts of New Guinea, a fantail is believed to be the ghost of a relative, and in Māori mythology, the fantail (pīwakawaka) woke the goddess of death by laughing as folk hero Maui was sneaking past, thereby causing his death.

Fantail nests consist of a cup of fine grasses, rootlets, or other vegetation, bound with spider webs, often with a trailing tail hanging from the base and normally placed in a tree fork. The female selects the nest site, and the pair both build the nest and take care of the brood. Females will pretend to be injured to lure away predators from the nest, and even outside the breeding season they fiercely defend their territories, attacking birds as big as a kookaburra (*Dacelo* spp.).

Most fantails are insectivorous, and often feed in mixed flocks containing other insectivores, catching insects they disturb and sometimes stealing food from others. They have rictal bristles around the beak to protect their face from insects while grabbing them in flight or gleaning them from foliage. When foraging, fantails flutter restlessly, with agile twists and loops in a bid to flush out food. Most species feed in the low-to-high canopy of trees, but some prefer deep undergrowth and shade, feeding close to the ground, while the willie wagtail is often seen hopping over grassy lawns in parks and gardens. Its habit of perching on the back of cattle or sheep has earned the willie wagtail the nickname of shepherd's companion.

Additions to the family

The silktails (*Lamprolia* spp.) are two unusual fantails endemic to Fiji. They feed in leaf litter and forage up and down branches. Being small, blue-black, and bright white on the rump and tail, they do not resemble other fantails. However, recent DNA studies have shown that they, and the closely related Drongo Fantail (*Chaetorhynchus papuensis*), do belong in the same family.

CERULEAN FLYCATCHER
Eutrichomyias rowleyi
This critically endangered fantail is endemic to the Indonesian island of Sangihe, where habitat destruction means it is at risk of extinction.

DRONGO FANTAIL
Chaetorhynchus papuensis
The glossy, blue-black Drongo Fantail lives in New Guinea montane forests and often feeds with other species,sounding the alarm at the sight of an approaching predator.

➤ Foes on the water
In a lake in Australia, a willie wagtail (far left) and a White-plumed Honeyeater (*Ptilotula penicillatus*) argue over a prime perch. These species will clash over territory and food.

◄ Rogue passenger
Drongos are famous for their aggression, especially when defending their nest or territory. Here, a Black Drongo (*Dicrurus macrocercus*) harasses a Crested Serpent Eagle (*Spilornis cheela*) in flight.

Drongos

Dicruridae

Members of this family of mostly black, long-tailed, strongly built birds are irrepressible and conspicuous songbirds found in Asia, Africa, and Australasia.

Drongos are perch-and-chase predators, watching from an elevated position before sallying off (often in swooping, twisting flight) to catch food in mid-air or from the ground. They are typically found in forest clearings, or in the canopy, where there is room to spot and chase prey, usually insects, such as grasshoppers, beetles, and flies. They have a strengthened palate for crushing the exoskeletons of large insects. Drongos will also catch small birds and even bats, using their strong bill to handle the squirming prey. They have bristles at the base of their bill, which help to protect their face.

These highly adaptable hunters attend bushfires to catch fleeing prey, often perched on an antelope or zebra's back for a better view. Drongos are renowned for their ability to imitate a flock member's alarm call so as to distract it and steal its food. They also imitate predators, and their loud, rambling, and inventive songs are the first to be heard at dawn and the last heard at dusk.

► Striking streamers
Drongo species are identified by tail shape and whether they have a crest. The Greater Racket-tailed Drongo (*Dicrurus paradiseus*) has distinctive tail streamers that can grow to 28 in (70 cm) long.

Order Passeriformes

Families 1

Species More than 20

Size range 7–26 in (18–65 cm) long

Distribution Australia, Africa, Asia, New Guinea

➤ In the pink
Australia's Pink Robin (*Petroica rodinogaster*) lives in rainforest and wet eucalyptus with a dense understory. Some birds migrate to more open areas after breeding.

This robin's call is a characteristic tik, tik, tik, while the song is a short, chattering trill

Only adult males have the bright pink breast and belly; females are mostly gray-brown

Order Passeriformes

Families 1

Species More than 50

Size range 4–9 in (10–22 cm) long

Distribution Australasia, Oceania

Australasian Robins

Petroicidae

These small, rotund, perch-and-pounce songbirds are so-named because their colorful breasts bear a resemblance to that of the much-adored, but unrelated, European robin.

SCARLET ROBIN
Petroica boodang
The Scarlet Robin is a common bird of bushland in eastern Australia. Pairs remain on territory all year, and some have been known to attempt five consecutive broods.

JACKY WINTER
Microeca fascinans
One of the few members of the family found in dry, open areas, the Jacky Winter is also one of the more widely distributed, being found throughout much of Australia.

Australasian robins feed on insects and worms that they gather in several ways, but notably by selecting an elevated perch and pouncing down to the ground to catch whatever they have spotted. They also hop along the ground or make sallying flights to catch insects on the wing. They have a distinctive tendency to cling to the sides of tree trunks and also to flick their wings and tail.

While many species typically occur in woodland, some have expanded to fill other habitats such as montane scrub and fast-flowing streams. For example, the Torrent Flyrobin (*Monachella muelleriana*) sallies for flying insects above the water in New Guinea and New Britain. Most Australasian robins make delicate cup-shaped nests constructed from fibers and bound with cobwebs. The Lemon-bellied Flyrobin (*Microeca flavigaster*) proportionally makes the smallest nest of any Australian bird—only one egg is laid, whereas two or three is the usual for this family.

Conservation success

A number of species in this family are rare or have small ranges. Remarkably, the Black Robin (*Petroica traversi*) of Chatham Island, New Zealand, was once the world's rarest bird, with only five individuals, including one female of breeding age. Conservation efforts have raised their total number to about 300 on two islands.

Orioles

Oriolidae

Often brightly colored, with enchanting, flutelike songs, orioles have long been prized by cultures around the world for their charm and beauty.

The true orioles are a diverse and often dazzling group of birds—their name comes from a Latin word meaning "golden." Most belong to the genus *Oriolus*, but the family also includes three figbirds (*Sphecotheres* spp.) in Asia and Australia, and four pitohuis (*Pitohui* spp.) in New Guinea. Despite similarities in looks and behavior, New World orioles belong to an unrelated family, the Icteridae.

Orioles are secretive birds, often hidden among the foliage of trees. They have wonderful, fluting songs, and when glimpsed are often revealed to have brightly colored plumage. Six brown-feathered orioles from Indonesi, Australia, and New Guinea, however, are a remarkable exception: they seem to mimic the larger friarbirds (*Philemon* spp.) in plumage, posture, and flight. The greater the size or weight disperity, the closer the plumage match: the Gray-collared Oriole (*Oriolus forsteni*) is almost identical to the Seram

Order Passeriformes

Families 1

Species More than 30

Size range 7–13 in (17.5–32 cm)

Distribution Europe, Africa, Asia, Australasia

◄ Asian oriole
This 1840s celebration of a Black-naped Oriole (*O. chinensis*) flying above a clematis is by Japanese woodblock master Utagawa Hiroshige.

▲ Insect hunters
Eurasian Golden Orioles (*O. oriolus*) hunt for insects, a source of food for nestlings, but berries and fruits form a greater part of the diet outside the breeding season.

◄ Green oriole
The Green Oriole (*Oriolus flavocinctus*) of eastern Indonesia and northern Australia lives in tropical evergreen rainforest, and parks and gardens.

AUSTRALASIAN FIGBIRD
Sphecotheres vieilloti
Like other figbirds, the Australasian figbird is very sociable, forming groups of more than 100 birds. The female is much duller than the brightly colored males.

HOODED PITOHUI
Pitohui dichrous
This poisonous oriole from New Guinea lives in family groups and has a diet of fruit and insects. Its black and orange plumage is mimicked by several other species.

BLACK-AND-CRIMSON ORIOLE
Oriolus consanguineus
While most orioles have yellow or green plumage, five Asian species have red feathering. The black-and-crimson oriole inhabits Indonesia and Malaysia.

" A mantis stalking a cicada is unaware of an oriole behind it. "

CHINESE PROVERB

Friarbird (*P. subcorniculatus*), which is about 75 percent heavier. The mimicry is thought to enable the orioles to escape attack by the larger friarbirds.

Food, faith, and toxicity

A 4,500-year-old relief carving on an ancient Egyptian tomb in Saqqarah shows people hunting orioles with nets attached to fig trees; and from Malta to Southeast Asia, orioles continue to be hunted for the pot, much to the anger of many conservationists. In the Altai Mountains of Central Asia, shamans use orioles in religious rituals.

The Hooded Pitohui (*Pitohui dichrous*), however, would exact revenge if eaten. People in New Guinea call it the "rubbish bird," having found that their tongue goes numb if they lick its feathers. When ornithologists also suffered burning mouths and nasal passages while handling the birds, they investigated and found that pitohui feathers and skin have strong concentrations of a neurotoxin previously known solely from poison-dart frogs of the tropical Americas. The hooded pitohui gets the toxin from its diet of melyrid beetles. Being poisonous may deter parasites, such as mites and ticks, or predators, or both.

The **Australasian figbird** will leave its **rainforest habitat** to search for **fig trees** in towns

Order Passeriformes

Families 1

Species More than 40

Size range 6¼in–3⅔ft (16–110cm) long

Distribution Australasia

➤ **Fascinating forms**
Two male King Birds-of-paradise (*Cicinnurus regius*) display above a female in Joseph Smit's illustration from 1883. At that time, skins of these birds were exported in large numbers to Europe as curiosities and for museums.

Birds-of-paradise

Paradisaeidae

Few birds capture the imagination like birds-of-paradise, with their ornate plumage, bold colors, and vivid displays. They mostly live on the island of New Guinea, where few people will ever see them.

The male birds-of-paradise exhibit some of the most outlandish plumage ever seen in the bird world; indeed, they show more structural variations of feathers than almost any other bird family. Flowing flank plumes, ribbonlike tails, bizarre crests, extended neck plumes, and wirelike feathers are all displayed alongside vivid colors of red, orange, pink, blue, and yellow. And almost every species has iridescent feathers that shimmer in the light.

It is not surprising, therefore, that these lavish adornments have entered deep into the cultures of the rural and forest-dwelling peoples of New Guinea. It is thought that bird-of-paradise feathers have been used in ceremonies, dances, and rituals for more than 10,000 years. The plumes are used in headdresses, there are dances that mimic the birds' displays, and their feathers are sometimes used as currency. Birds-of-paradise

➤ **Putting on a show**
A male Raggiana Bird-of-paradise (*Paradisaea raggiana*) raises his splendid feathers in a full courtship display. Some males of this species have been known to display in the same lek for 20 consecutive years.

feature in many local myths and legends and are the national emblem of Papua New Guinea and the logo of the national airline.

Impressive dances

The traditional dances of New Guinea peoples reflect the lavish courtship displays of the birds themselves. There is an extraordinary diversity of avian choreography, from jumping to running, ruffling, flapping, and bowing. The display of the Queen Carola's Parotia (*Parotia carolae*) contains 58 distinct elements and several species, such as the Blue Bird-of-paradise (*Paradisornis rudolphi*), hang upside down, ruffling their feathers.

All displays are accompanied by complicated call notes, many of them loud and with a wailing tone, which alert females to the male's location. The five manucode species (*Manucodia* spp.) have a greatly elongated windpipe (trachea), which coils around the pectoral muscles and lowers the pitch of their call, making it carry further through the forest.

Birds-of-paradise also display in a variety of social arrangements—some males perform alone while others display in loose aggregations known as a lek. Display dances may be on the ground, just above the ground, or in the treetops. Some birds prepare a court for their display, keeping the forest floor clean and removing leaves from vines before beginning their performance. The best males invariably occupy the top display court, often with an elevated perch, and males fight vigorously for a performance slot in order to have a chance of success. It can take young

> "As well as wearing a fantastic costume he is also, in effect, a trampoline artist."
>
> DAVID ATTENBOROUGH, *The Life of Birds*, 1998

◄ Bright but elusive
The extraordinary Wilson's Bird-of-paradise (*Diphyllodes respublica*) is found only in the forested hills of Waigeo and Batanta, two islands northwest of New Guinea.

male birds-of-paradise several years to learn the moves for their display dances and some seven years to grow their elaborate adult plumes.

The same but different

Despite their great variation, birds-of-paradise have several unifying features. They are generally quite large, vigorous songbirds with powerful legs and feet, and they can grasp and hold food to their bill. The bill, which may be either straight or downcurved, is large and powerful. Most birds-of-paradise hop on the ground and fly with a mix of flapping and gliding.

Some species of bird-of-paradise form permanent pairs that share parental responsibilities. In most species, however, the males mate with several females during the breeding season and the females incubate their eggs and care for their young alone. The females build cup-shaped nests from vines, the stems of orchids, ferns, moss, and leaves. Occasionally, a shed snakeskin is added to the nest, which may be an ingenious way the bird protects its nest from predators.

Feathered headdresses
Performers at a Sing Sing, a festival-like gathering of peoples in Papua New Guinea, wear traditional clothing that features the red and orange plumes from the Raggiana bird-of-paradise. Plumes are often traded or borrowed for these social events.

BROWN SICKLEBILL
Epimachus meyeri
Found in the central highlands of New Guinea, this species has a curved bill that is ideal for probing for invertebrates, although it also eats fruit.

KING OF SAXONY BIRD-OF-PARADISE
Pteridophora alberti
With astonishing neck plumes that are held up and apart in display, this unique species makes an extraordinary call that has been likened to radio static.

BLUE BIRD-OF-PARADISE
Paradisornis rudolphi
This large and uncommon species sings from a high perch and then descends to lower perches where it hangs upside down to perform its display.

Order Passeriformes

Families 1

Species More than 100

Size range 3½–12 in (9–30 cm) long

Distribution Africa, Asia, Australasia, Oceania

The head and crest are black on both color morphs

Spreading the wings wide in flight gives the paradise flycatcher great control

The central tail feathers project up to 12 in (30 cm) beyond the others

▲ Polymorphic plumage
The male Indian Paradise Flycatcher (*T. paradisi*) has two color morphs: one is reddish-brown with gray-white underparts, as here, and the other is mostly white.

Monarchs

Monarchidae

The **tail feathers** of male **paradise flycatchers** can be more than **double** the bird's **overall length**

This family of flycatchers comprises many brightly colored species, some with spectacularly long central tail feathers. Many have a wide, flat bill that helps them catch insects in midair.

The monarchs are a large family with a wide range, but they have many features in common. These birds are typically small- or medium-sized, with a slender body shape and upright posture. Their feeding strategy is also fairly consistent. Monarchs normally remain motionless for long periods before sallying at flying insects or gleaning crawling insects from vegetation with their broad-based, flattened bill.

A monarch's plumage is generally dark above and light below, and more than a third of species—ranging from Indonesia's White-naped Monarch (*Carterornis pileatus*) to Australia's Pied Monarch (*Arses kaupi*)—are black and white. Others exhibit reddish feathers, and Black-naped Monarch males (*Hypothymis azurea*) are vivid blue. Many family members have a crest, but it is only the 17 species of paradise flycatcher (*Terpsiphone* spp.) whose males boast very long tail feathers. In flight, the extravagant appendage floats behind like a streamer rippling in the breeze. In some places, paradise flycatchers generate local income by attracting birdwatching visitors. For example, many tourists visit the reserve on La Digue, Seychelles. It was established in 1982 to provide a protected breeding habitat for the Seychelles Paradise Flycatcher (*T. corvina*), and numbers have since recovered.

◄ Beware magpie-larks
This aggressive bird is found across most of Australia. It readily attacks its own reflection in windows, and is bold enough to chase birds as large as eagles, and even people who approach its nest.

Trusted spirits

Monarchs largely live in wooded habitats, from savannas and mangroves to dense tropical rainforest. Three-quarters of species inhabit single islands or island groups. Such a small domain means that they are susceptible to habitat loss and to non-native predators like rats. Some monarchs, such as shrikebills (*Clytorhynchus* spp.) and flycatchers in the *Myiagra* genus, inhabit Pacific islands. In Hawaii, elepaios (*Chasiempis* spp.) were revered as guardian spirits of canoe-makers. Each time a koa tree was chopped down to make a new canoe, the craftspeople monitored the elepaio's behavior. If the bird pecked at the wood in search of insects, it meant that the tree was too soft to be seaworthy.

MAGPIE-LARK
Grallina cyanoleuca
Distinctively black and white and often seen perching prominently, the Magpie-lark is one of Australia's most familiar birds. Although most monarchs feed in trees, this species usually feeds on the ground.

The black-naped monarch can become caught in the webs of large spiders

Shared care
Breeding pairs of Asia's handsome black-naped monarchs share parental responsibilities, including incubating the eggs and feeding their young.

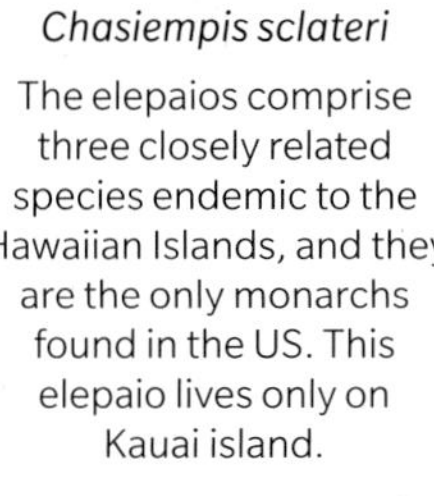

KAUAI ELEPAIO
Chasiempis sclateri
The elepaios comprise three closely related species endemic to the Hawaiian Islands, and they are the only monarchs found in the US. This elepaio lives only on Kauai island.

RENNELL SHRIKEBILL
Clytorhynchus hamlini
This species is found on Rennell Island in the Solomon Islands. Unusually, shrikebills have a long, hook-tipped bill, which is perfect for foraging.

Order Passeriformes

Families 1

Species More than 130

Size range 8½–26 in (21.5–65 cm) long

Distribution Worldwide except Antarctica

◄ **Trick of the light**
North America's Blue Jay (*Cyanocitta cristata*) has stunning blue plumage. However, its feathers in fact contain no blue pigments and only appear to be different shades of blue because of the way they scatter light.

▲ **Mountain magic**
The Inca Jay (*Cyanocorax yncas*) lives in the forests of the Andes. This dazzling bird can produce many distinctive calls, including one like an alarm bell.

Crows, Jays, and Magpies

Corvidae

Studies have found that **ravens remember** the faces of **people** who have **treated them badly**

Renowned for their curiosity and ingenuity, crows, jays, and magpies play a special role in the folklore and superstitions of many cultures around the world.

Known as corvids, the members of the crow family have many traits that set them apart from most songbirds. They do not have tuneful songs, although many species are vocally talented. They use a wide range of calls to communicate with one another and possess impressive powers of mimicry. Most corvids are also much larger than typical songbirds, and the Common Raven (*Corvus corax*) and Thick-billed Raven (*C. crassirostris*) are the largest passerines of all, weighing about 3.3 lb (1.5 kg). Most species with "crow" or "raven" in their name are monochrome (usually plain black), but some jays and magpies, especially those that live in forest habitats, have vivid coloration. Some jays also have impressive crests.

Most corvids are generalist feeders and are noted for their intelligence and curiosity. Some have learned novel ways to find food in the wild, such as a flock of Carrion Crows (*C. corone*) in Japan. They place nuts on a road intersection when traffic is stationary at the lights, and then collect the kernels on the next traffic light cycle after passing vehicles have cracked the shells. Studies on captive northern ravens and New

Caledonian Crows (*C. moneduloides*) have revealed the birds' impressive powers of learning, planning, and innovation, as they solved complex physical puzzles, created and used tools, and demonstrated teamwork, in order to access edible treats.

Many corvids have strong social bonds, and Eurasian magpies (*Pica pica*) are among the few animals to have passed the "mirror test." This means they understand that the image they see in a mirror is a reflection of themselves and not another bird. They show awareness that other birds have their own separate experience of the world (for example, that a treat may be visible to one bird but not to another that is looking from another angle). In wild corvid populations, knowledge and skills are passed down from

▲ True crows
A widespread Eurasian species, the Hooded Crow (*Corvus cornix*) lives on farmland and wilder open countryside, as well as around towns and villages.

◄ Odin's ravens
The Norse god Odin has two talking raven familiars, who travel the world to collect information. Huginn (meaning "Thought") and Muninn ("Memory") are often depicted perched on his shoulders, as in this 18th-century Icelandic manuscript .

The robust head and strong multipurpose bill, useful for digging in soil, is typical of a "true crow" (genus *Corvus*)

Strong wings beat in a steady "rowing" action with little or no gliding

parent to offspring. This is known as cultural transmission, and it means each population may have its own rich and complex culture.

Masters of memory

Clark's Nutcracker (*Nucifraga columbiana*), a small North American corvid, has a different but equally impressive mental skill. It is a "scatter hoarder," hiding huge numbers of pine seeds in thousands of caches throughout the fall. It survives the snowy winter by digging up these seeds, managing to remember where each cache is buried. Like other corvids, its brain has a high density of neurons, and is particularly suited to this memory task. The Canada Jay (*Perisoreus canadensis*) hoards nuts similarly, but faces competition from a rival corvid, Steller's Jay (*Cyanocitta stelleri*), which follows Canada jays around in hope of stealing their stashes. The Canada jay has also learned to hang around human campsites to scavenge food and is often affectionately nicknamed "Whisky Jack" after the shapeshifting creator and trickster Wisakedjak in Cree and Algonquin folklore. This jay was declared national bird of Canada in 2016—an appropriate choice as it is widespread in

Red disk in the bird's mouth represents the light stolen by Raven

▲ Ceremonial rattle
Wooden raven rattles, like this 19th-century example, were used in shamanic dances and healing ceremonies by the Haida people of western Canada. It is filled with stones or seeds.

CLARK'S NUTCRACKER
Nucifraga columbiana
This stocky bird holds pine cones in one foot while using its long pointed bill to extract seeds from cones. It will also eat insects, eggs, and young birds.

HOUSE CROW
Corvus splendens
In southern Asia, this crow is a familiar town-dweller and is typically curious and opportunistic. Stowaways on ships have established populations in Hong Kong and Florida and elsewhere.

ALPINE CHOUGH
Pyrrhocorax graculus
A true high-altitude specialist, the acrobatic Alpine chough will breed high in the Himalaya, and has been seen foraging at 27,000 ft (8,200 m) by climbers on Everest.

◄ Repelling an intruder
The larger corvids are highly intolerant of any birds of prey in their airspace and will mob them mercilessly. Here an American Crow (*Corvus brachyrhynchos*) chases off a young red-tailed hawk.

Canada, but its range barely extends into the northern US. It is the only Canadian bird regularly referred to by an indigenous name.

Many cultures celebrate the cleverness and mischievous spirit of crows and ravens in folk tales and mythology. A common theme in First Nation tales in Australia and North America is that the crow was once white, but scorched its feathers black through some misadventure—or as punishment for theft or trickery. One myth told by the Haida and Tinglit peoples of western Canada describes how Raven (a trickster character in their myths) transforms into a baby to steal light from the Sky Chief. He escapes with light, but drops half as the stars and the moon, and releases the rest as the sun. Ravens also make appearances as messengers in the Bible and the Quran; the Biblical raven was the first bird that Noah sent out from the Ark in search of land. Chinese mythology associates the crow with the sun, with three-legged "sun crows" described as guiding the sun on its journey across the sky.

" But the Raven, sitting lonely on the placid bust, spoke only That one word, as if his soul in that one word he did outpour. "

EDGAR ALLAN POE, *"The Raven" 1845*

Avoiding bad luck

The Eurasian and Black-billed Magpies (*Pica pica* and *P. hudsonia*) are strongly associated with superstitious beliefs in some cultures. In the UK, for example, a lone magpie brings bad luck unless it is "saluted"—the viewer must say, "Good morning, Mr. Magpie," and ask after the health of the magpie's "wife." Fortunately, it is rare to see a solo magpie as these birds are quite gregarious, and in the breeding season males are diligent "mate guards," following their partner closely wherever she goes in readiness to chase off any potential rival. The widespread European belief that magpies are inveterate thieves of sparkly trinkets was proved untrue in a 2014 UK study that found that the birds actually tend to be wary of unfamiliar shiny objects, although corvids in general are curious and will (cautiously) explore new places in hope of finding new feeding opportunities.

CLEVER CROWS

The Hawaiian Crow (*Corvus hawaiiensis*) uses a stick to probe a tree crevice that is too deep and narrow to be explored with its bill when looking for food. This corvid, extinct in the wild, is closely related to the New Caledonian crow, another highly intelligent island species. The New Caledonian crow crafts its own tools, removing leaves from a stick and then bending it into a hook shape.

HAWAIIAN CROW

Some crow species are common across many countries, and even thrive in heavily human-modified landscapes thanks to their adaptable ways. However, some specialized island-dwelling species, such as the Mariana Crow (*Corvus kubaryi*) of Rota, and the Flores Crow (*C. florensis*) of Flores island in Indonesia, are threatened, as is the Island Scrub Jay (*Aphelocoma insularis*) of Santa Cruz—which has the distinction of being the only North American land bird that is endemic to a single island. The Azure-winged Magpie (*Cyanopica cyanus*) in eastern Asia is extremely similar to the Iberian Magpie (*C. cooki*) in Spain and Portugal, despite being separated by about 5,000 miles (8,000 km). They are probably derived from a common ancestor, whose population was split during the last ice age.

➤ Tengu mask
In the Japanese Shinto religion, tengu are supernatural entities which may take on monstrous birdlike forms. This 18th- or 19th-century wooden mask represents a crow tengu.

An exaggerated bill is a feature in many avian tengu artworks

Another corvid extremist is Eurasia's Alpine Chough (*Pyrrhocorax graculus*)—those living in the Himalaya probably breed at higher altitudes than any other birds. Any mountain camp or ski lodge in the right area will attract flocks of these smallish, yellow-billed crows, which show the same opportunistic confidence as their lowland cousins. However, corvids can themselves be exploited, and some species are regularly hosts to brood parasites, unwittingly bringing up the chicks of other species. For example, Great Spotted Cuckoos (*Clamator glandarius*) target Eurasian magpies, while in South America the dazzling Inca Jay (*Cyanocorax yncas*) falls victim to the Giant Cowbird (*Molothrus oryzivorus*).

▼ The thief
In European folklore, magpies often stand accused of stealing jewelry. This 19th-century watercolor by Clementina Margaret Hull shows a magpie stealing a necklace from a nobleman's palace.

➤ Seedhead acrobats
Black-capped chickadees are typical of their family for their acrobatic ability to reach anything edible, from a seed in a flowerhead to a caterpillar on a treetop leaf.

A short conical bill is ideal for husking seeds and grabbing insects

Extremely strong feet give the bird at tight grip—it can perch upright or hang upside down

Relatively short rounded wings help the bird maneuver through cluttered habitats, such as woodland and forest

Tits and Chickadees

Paridae

These distinctive small, restless songbirds have strong legs and feet for feeding acrobatically in trees and shrubs. Bold, noisy, and inquisitive, they are a popular sight at feeders in many countries.

Tits and chickadees form a highly engaging, yet remarkably uniform family. Virtually all members are small, highly active tree-dwelling birds. They all have specialized leg muscles and strong feet to help them grip tightly, allowing them to feed in most parts of a tree—from the smallest treetop twigs (often hanging upside down) to the trunk, and on the ground beneath. They have a short bill adapted to a varied diet, which in temperate regions generally switches from insects, especially caterpillars, in the breeding season to seeds and nuts in the winter. Many tits and chickadees have bold patches, such as black caps and throats, and some are brightly colored. Quite a few have crests. They have soft plumage with minimal sexual dimorphism—at least to human eyes.

Tits and chickadees are among the world's most popular birds, occurring in abundance in North America and Eurasia, where people in

Order Passeriformes

Families 1

Species More than 60

Size range 3½–8 in (8.5–21 cm) long

Distribution North America, Eurasia, Africa

their millions put out hanging feeders and enjoy the antics of these birds. In Europe, Eurasian Blue Tits (*Cyanistes caeruleus*) and Great Tits (*Parus major*) are abundant, while in North America, Black-capped Chickadees (*Poecile atricapillus*) are found throughout much of the continent, giving way to Carolina Chickadees (*P. carolinensis*) in the southeastern US, and Mountain Chickadees (*P. gambeli*) in the Rocky Mountains to the west. Over the past 70 years, the range of the Tufted Titmouse (*Baeolophus bicolor*) has expanded north from the southeastern US into New England and southern Canada, helped in part by the provision of food in feeders.

Cache management

One of the intriguing characteristics of the family, adopted by some but not all species, is the ability to store food in caches. This might be for a few hours or days, or even months in some cases. Items, such as nuts, but also invertebrates (sometimes beheaded to paralyze them) are hidden away in leaf litter, in cracks in the bark, or under moss to be retrieved when required. Some birds store prodigiously: Willow Tits (*P. montanus*) may cache up to 150,000 items during one fall. The birds find these caches from memory, and in the case of the Marsh Tit (*P. palustris*), its hippocampus (the part of the brain that deals with spatial cognition) is 31 percent larger than that of the non-storing great tit. It usually retrieves its oldest caches first. Another food-storer is the Gray-headed Chickadee (*P. cinctus*), a bird that can tolerate outdoor temperatures of -49°F(-45°C) in the Arctic. It has exceptionally dense plumage and can lower its body temperature by at least 18°F (10°C) at night to conserve heat. Also known as the Siberian tit, it is the only tit found in both the Old and New World.

Although tits and chickadees are mostly associated with Eurasia and North America, 15 species occur in sub-Saharan Africa. Many of these have black plumage, but often have significant white in the wings. However, the Red-throated Tit (*Melaniparus fringillinus*) of Kenya and Tanzania, as its name suggests, has a russet head and chest. Like their northern cousins, these species eat small invertebrates and larvae, as well as seeds and fruit. They do not store food, and they tend to have small clutches of 3–4 eggs.

The majority of tits and chickadees are socially monogamous and hold territories that they primarily defend by song. Scientists have recently

Great tits have been known to catch and eat hibernating bats

➤ Crested charmer
The Crested Tit (*Lophophanes cristatus*) is found in forests across much of Europe. In the north of the continent, it is strongly associated with pine forests, but can also be found in deciduous forests in southern Europe.

▲ Taking the crown
The magnificent Sultan Tit (*Melanochlora sultanea*) is by far the largest of the tits, reaching 8 in (21 cm) in length, almost twice the size of a great tit.

discovered that male great tits that live in areas with dense foliage cover sing lower-pitched songs than those birds living in more open habitats. This is because lower-pitched sounds travel better and do not become muffled in the undergrowth. They have also found that great tits in cities sing at a higher pitch in order to be heard above the rumble of traffic, another example of birds of this family adapting to human-created environments.

Colorful clues

The plumage of tits and chickadees holds clues as to an individual bird's health and abilities. For example, the feathers in the crown of male Eurasian blue tits reflect ultraviolet light. This feature varies between individuals and females prefer males with the brightest crowns. Meanwhile, the yellowness of a blue or great tit's breast has been correlated to its ability to obtain carotenoids (a yellow or red pigment) from its prey, caterpillars. A bright coloration could indicate overall fitness, a good habitat or territory, and enhanced foraging ability.

Tits and chickadees invariably nest in holes, usually in trees. There is an intriguing dichotomy in the family: food-hoarding species excavate their own nest-holes, usually in old, rotting

> "Nor will it easier be—nay, not a whit—
> To keep from your domain the greedy Tit.
> Small is the naughty Fowl, yet it can wreak
> No small Destruction with its claws and beak."

FATHER JEAN IMBERDIS, *Papyrus*, 1693

SOUTHERN BLACK-TIT
Melaniparus niger
Found in southern Africa, this is a bird of savanna and light woodland. Its diet includes fruit, nectar, wasps,and other insects.

VARIED TIT
Sittiparus varius
A common Japanese species, this bird used to be caught and used at shrines, where it would pluck fortune-telling slips declaring the fate of visitors.

BRIDLED TITMOUSE
Baeolophus wollweberi
This woodland bird from Mexico and the southern US often forms the nucleus of mixed-species foraging flocks. It is often first to give an alarm call when a predator is near.

wood, while the rest simply appropriate any suitable cavity. Within the hole, the female builds a cup-shaped nest, usually out of moss, with varying amounts of grass, feathers, and leaves. Eurasian blue tits, and probably other species, add aromatic plant fragments, which can help kill bacteria and help the chicks grow faster. The female Eurasian blue tit lays an average clutch of 11 eggs and sometimes as many as 16. During nest-building and incubation, the male brings food to the female. The southern black-tit is unusual for having helpers (usually the young from previous broods) at the nest to assist with raising chicks. The nest can be a frantic place when the eggs hatch. In several species, both male and female provision their large broods with enormous numbers of caterpillars, up to 1,000 a day. The adults often break the jaws of larger caterpillars so that they do not injure the mouths of the nestlings.

The crown of the male bird reflects large amounts of ultraviolet light

◀ Blue wonder
The Eurasian blue tit is one of Europe's most abundant species. Bold and inquisitive, it is typically found in oak woodland, but has also become a familiar sight in parks and gardens.

Tits are often highly sociable outside the breeding season and may form the nucleus of winter foraging flocks of multiple species. Some individuals wander widely during the day, while in the case of marsh tits and Carolina chickadees, a pair or group lives within its own winter territory and only joins flocks passing through.

▼ Beating the freeze
The willow tit can live all year beyond the Arctic circle. At night, it is able to dig its way into the snow to sleep within a burrow that is slightly warmer than outside.

GREATER HOOPOE-LARK
Alaemon alaudipes

This desert lark of North Africa, the Middle East, and Asia has long legs and a long body. It reveals black-and-white bands across the wings as it flies. It will not breed in very dry years.

RUFOUS-NAPED LARK
Mirafra africana

Common in African savannas, this lark has more than 20 subspecies. It has a reddish-brown neck and flight feathers. Its short, blunt crest is typical of many lark species.

◄ Flying joy
A Eurasian skylark sings high above a landscape in this illustration from 1884 inspired by an Italian Renaissance painting. Skylarks sing to proclaim ownership of a territory and to attract a mate.

Larks

Alaudidae

Some larks have a crest or horns, but most are small, unassuming birds. Many species, however, have an exquisite song—a long and complex melody made up of warbling notes, trills, and whistles.

Order Passeriformes

Families 1

Species More than 90

Size range 5–10 in (12–24 cm) long

Distribution Worldwide except Antarctica

Larks are ground-living songbirds that have slim legs and long hind claws. They are found in many open habitats, such as deserts, farmland, exposed moorlands, even on mountain peaks. Modern farming can be harmful to them, and in western Europe, for example, populations of the Eurasian Skylark (*Alauda arvensis*) have fallen. While most species avoid trees, others perch on bushes, and some, such as the Woodlark (*Lullula arborea*), sing or launch their circling song flights from high trees.

Endless variety

Despite modest brown plumage, some larks have evolved distinct local forms. The Desert Lark (*Ammomanes deserti*) has 22 subspecies, while the Crested Lark (*Galerida cristata*) has 33. Their plumage may be darker, paler, redder, or grayer to help them blend in with local soils.

Larks often appear in everyday speech in some cultures. For example, "up with the lark" alludes to their activity at daybreak (as mentioned by Chaucer and Shakespeare). And they are seen as happy, joyful, carefree birds, reflected in phrases such as "larking about." More unfortunate is their connection with food. Larks' tongues were a delicacy in ancient Rome despite their minute size—one recipe required 1,000 tongues. And large numbers of larks are still eaten in some Mediterranean countries. In England, skylarks used to be caught by children with flickering mirrors, and cooked, bones and all, in pies.

The **skylark's** rapid **flow of song** during its high, hovering flight may **last unbroken** for **30 minutes**

Tiny black "horns," shorter in other subspecies, are most obvious on the male in spring

The broad black mask distinguishes the Caucasian horned lark from the other subspecies

➤ Masked singer
The Horned Lark (*Eremophila alpestris*) has more than 40 subspecies. This one, a Caucasian Horned Lark (*E. a. penicillata*), is found from Turkey to Iran.

" How the blithe Lark
runs up the golden stair
That leans thro' cloudy gates
from Heaven to Earth, "

FREDERICK TENNYSON, *The Skylark*, 1854

Order Passeriformes

Families 1

Species More than 60

Size range 5–8 in (12–21 cm) long

Distribution Worldwide

Wagtails and Pipits

Motacillidae

These elegant, characterful birds are found in grassland habitats around the world, although some species favor woodlands and river valleys. They live on all seven of Earth's continents.

YELLOW-THROATED LONGCLAW
Macronyx croceus
This is the most widespread of the eight longclaw species, all of which live in sub-Saharan Africa's grasslands. It has bright yellow or orange plumage on the underside.

Wagtails and pipits are small, slender birds with a wide range of plumage patterns. While wagtails often sport striking combinations of black and white, or vibrant shades of yellow, orange, or green, pipits usually come streaked in subtle shades of brown. All share white outer feathers on their narrow and variably long tails.

Tail twitchers

Most species twitch their tails constantly as they walk, which is especially noticeable in wagtails due to the length of the tail. This behavior gives the family its scientific name ("motacilla," the Latin for wagtail, derives from a word that means "to move"). The reason they do this is uncertain: it could be a tool for flushing out prey, a social signal, or an indication of a bird's state of alertness. Another, less conspicuous feature is the very long hind claw on the backward-pointing toe, which offers these birds greater stability and grip as they maneuver through uneven grasses on a forest floor.

Wagtails and pipits are primarily insect eaters, although they will also prey on spiders, worms, crustaceans, and even some small vertebrates depending on food availability. In colder conditions, when insects become hard to find, they also consume seeds and berries.

" Little trotty wagtail, he went in the rain,
And tittering, tottering sideways
he ne'er got straight again. "

JOHN CLARE, "Little Trotty Wagtail," 1849

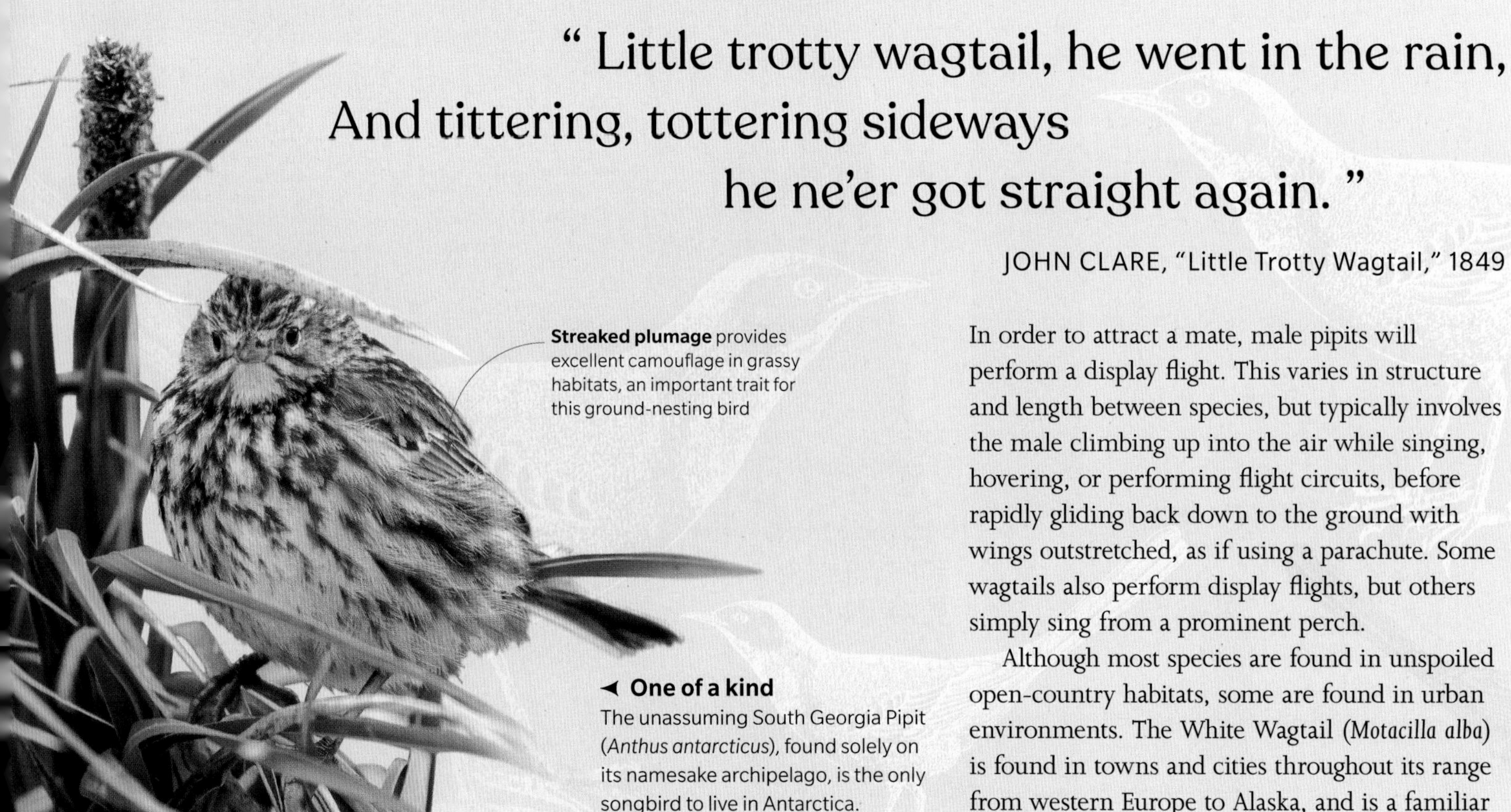

Streaked plumage provides excellent camouflage in grassy habitats, an important trait for this ground-nesting bird

◄ One of a kind
The unassuming South Georgia Pipit (*Anthus antarcticus*), found solely on its namesake archipelago, is the only songbird to live in Antarctica.

In order to attract a mate, male pipits will perform a display flight. This varies in structure and length between species, but typically involves the male climbing up into the air while singing, hovering, or performing flight circuits, before rapidly gliding back down to the ground with wings outstretched, as if using a parachute. Some wagtails also perform display flights, but others simply sing from a prominent perch.

Although most species are found in unspoiled open-country habitats, some are found in urban environments. The White Wagtail (*Motacilla alba*) is found in towns and cities throughout its range from western Europe to Alaska, and is a familiar

⮝ Odd bird out
The Forest Wagtail (*Dendronanthus indicus*) has distinctive black and white wing bars. It is found in forested areas of East and Southeast Asia, and is the only wagtail that nests in trees.

sight in parks and gardens. In winter, large and conspicuous roosts are found in city centers, taking advantage of the urban warmth.

Cultural contrasts

This close coexistence with humans means that wagtails have a rich cultural history across their ranges. In a creation myth of the Ainu peoples of Japan, a wagtail creates the first land. Romani mythology sees them as a symbol of good fortune, and the Gypsy Lore Society used to have a wagtail as its emblem. On the other hand, a Gaelic superstition warns that the sight of a Gray Wagtail (*M. cinerea*) around the house is a sign of bad weather to come, while one appearing between a person and their house means that they will be evicted soon.

➤ Wagtail and lotus
This early 20th-century print by Ohara Koson, a master of "bird-and-flower" art, depicts a Japanese Wagtail (*M. grandis*), a large species native to Japan and Korea.

Order Passeriformes

Families 3

Species More than 170

Size range 3½–8 in (9–20 cm) long

Distribution Europe, Asia, Africa, North America

The lower mandible is a pinkish color, whereas the upper one is blackish-brown

Leaf, Reed, and Bush Warblers

Phylloscopidae, Acrocephalidae, Cettiidae

Warblers are vocally diverse, and their songs range from simple trills to sophisticated, musical, warbling phrases. They are slender birds with a slim bill, and most have understated rather than showy plumage.

Across most of their range, a few species of warblers make themselves known by their distinctive song. In Europe, for example, the Common Chiffchaff (*Phylloscopus collybita*), named for its repetitive song, may be heard in woodlands, parks, and gardens. Most species of warblers, however, are less well-known, occupying specialized habitats or being secretive in their behavior. Many migrate considerable distances, and a tiny minority sometimes wander off course. For example, several Siberian species move southeast in the fall, and a few accidentally make a "reverse migration," turning up in Western Europe. Although exciting for birdwatchers, this phenomenon can make identification more challenging.

Leaf warblers typically occupy woodlands or bushy places. Many species are green above and lighter beneath, usually with a pale line over the eye and a dark eyestripe. Some attract attention by their strong head patterns and pale wingbars. They feed on insects and spiders from leaves, slipping easily through foliage, sometimes hovering briefly, and often calling as they move around. Their song can

▲ **Quick-fire delivery**
The aquatic warbler sings alternating rapid, scratchy trills and musical phrases. It depends on a habitat that has largely disappeared across its range: marshy vegetation, up to 32 in (80 cm) high.

JAPANESE FACIAL

Common in much of Japan, where it is called the uguisu, the Japanese Bush Warbler (*Horornis diphone*) is a popular songbird, and it has been kept in cages for 1,200 years or more. Its droppings contain an enzyme that is used in facial creams, to whiten the skin and reduce wrinkles. Known as the "geisha facial," or *uguisu no fun*, this treatment was originally used by geishas to remove their makeup.

WOODCUT, KITIGAWA UTAMARO, C.1800

◄ **Reed quartet**
This 18th-century illustration shows two Aquatic Warblers (*Acrocephalus paludicola*) left, a Sedge Warbler (*A. schoenobaenus*) top right, and a Common Reed Warbler (*A. scirpaceus*) by its nest.

be bright and lyrical, although some species are less musical, and offer a simple rapid trill. Reed warblers are usually a little stockier and browner. Most are small, but species such as the Great Reed Warbler (*Acrocephalus arundinaceus*) are as large as a medium-sized thrush. They may sing from exposed perches—in rhythmic and musical, or rapid and scratchy phrases—but otherwise they remain mostly out of sight, deep down in wet vegetation. Their strong feet enable them to move through dense vertical stems. Bush warblers are largely brown or reddish-brown, and can be difficult to spot because they feed within thickets and marsh vegetation.

Spreading joy

In Western cultures, warblers are less familiar in stories and legends than songbirds such as the Common Nightingale (*Luscinia megarhynchos*), but in some Asian cultures, warblers are commonly associated with good fortune and new beginnings. For example, migrants that reappear one day, singing from the very same tree as the year before, are potent signs of the renewal of spring—just like a cuckoo.

◄ Fresh and bright
Of the Western European warblers in the genus *Phylloscopus*, the Wood Warbler (*P. sibilatrix*) has the brightest face and wing feather fringes. It alternates between two song types: a sweet whistle and a silvery trill.

" This was the best of May–the small brown birds
Wisely reiterating endlessly
What no man learnt yet, in or out of school. "

EDWARD THOMAS, "Sedge-Warblers," 1915

ARCTIC WARBLER
Phylloscopus borealis
This small greenish and whitish warbler, with a long eyebrow stripe and pale wingbar, breeds in northern birch woods and wet bushy places in Eurasia. It migrates to Southeast Asia in winter.

ICTERINE WARBLER
Hippolais icterina
Heavier and stronger billed than species in the genus *Phylloscopus*, this warbler likes higher treetops, standing tall within a wooded area, and has a fast imitative song.

CETTI'S WARBLER
Cettia cetti
This reddish-brown bush warbler can be located by its sharp call note and outbursts of musical song. It forages in dense cover near water for insects, spiders, snails, and some seeds.

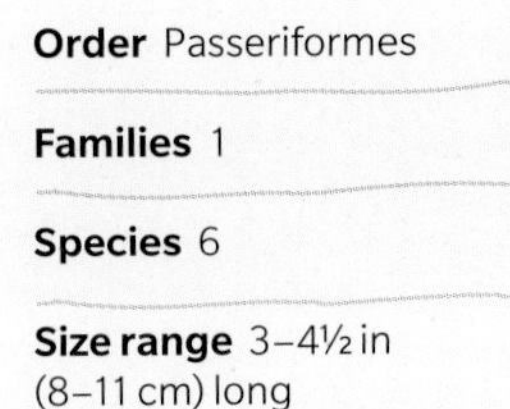

Order Passeriformes

Families 1

Species 6

Size range 3–4½ in (8–11 cm) long

Distribution Europe, Asia, North America, Africa

➤ Constant fidget
Like all kinglets, the North American Ruby-crowned Kinglet (*Corthylio calendula*) is very active, flitting between one perch and the next, in search of food.

Kinglets

Regulidae

Kinglets, or "crests," are some of the smallest songbirds within their range in North America and Eurasia, and they can often be located by their high-pitched voices.

Kinglets get their common name from the narrow "crown" of bright color on top of the male's head. It is usually folded into a thin stripe, edged with black, that lays flat. However, when kinglets sing and display, these crest feathers open up like a fan, to reveal the "crown" that defines the "king."

Frantic foragers

Smaller than most warblers, kinglets are busy birds, feeding in an acrobatic fashion, often hanging upside down, or hovering to capture insect prey, high among the thinnest twigs and lightest foliage. They also explore denser thickets, and especially the heavy fronds of evergreens. Within a mixed wood, Eurasian species frequently seek out a conifer, holly bush, or growths of ivy on mature oak trees. Often almost oblivious to human presence, kinglets can be observed closely. However, their thin calls and short songs can be "missed" by people who lose high-pitch hearing with age, thereby affecting field workers who survey bird populations. Nevertheless, song is still the best way to locate kinglets in summer, and also a good way to separate the similar-looking Goldcrest (*Regulus regulus*) from the Common Firecrest (*R. ignicapilla*)—the former having a distinct rhythmic song, the latter a more even trill.

Broad wings produce a fast, blurry, flitting action and an occasional whirring hover

▲ Tiny but mighty
The Eurasian goldcrest weighs barely more than a teaspoon of sugar, yet northern populations migrate thousands of miles to avoid fierce winter conditions, regularly crossing the North Sea.

◄ Sermon to the birds
This 20th-century painting shows the friar known as St. Francis of Assisi giving a sermon to a gathering of birds. According to the story, the birds, including Eurasian blackcaps and garden warblers, responded by singing to him.

Sylviid Babblers

Sylviidae

Once part of a large family of small, slender birds called Old World warblers, this new, much smaller grouping is the result of DNA investigations.

Birds in the *Sylvia* genus have beautiful songs. In England, the Eurasian Blackcap (*S. atricapilla*) is nicknamed the "northern nightingale," while the French composer Oliver Messiaen wrote music based on its song and that of the Garden Warbler (*S. borin*). Birds in the *Curruca* genus have more scratchy vocalizations and many have song flights.

Primarily insect-eaters, the Sylviid babblers will feed on fruit in the winter, eating small berries or pecking at larger fruits, such as dates and figs. Fruit is also an important food before migration. Asian and African Desert Warblers (*C. nana* and *C. deserti*) often feed alongside wheatears (*Oenanthe* spp.) from the chats and flycatchers family, which act as sentinels to spy out predators. Most species of Sylviid babblers are migratory although some populations are resident. The ancient Greek philosopher Aristotle thought that summer breeding garden warblers changed into blackcaps, which spend the winter in the Mediterranean.

Evolution in action

The Eurasian blackcap shows a migratory divide in Central Europe, with western birds moving southwest toward West Africa, and eastern birds passing through the Middle East as far as East Africa. They also began to establish a new migration route in the 1960s from Germany, northeast to Britain in winter. This migratory population now breeds earlier than others and is becoming genetically distinct with different wing and beak shapes, as well as beak color.

Order Passeriformes

Families 1

Species More than 30

Size range 4½–7½ in (11.5–19 cm) long

Distribution Europe, Africa, Asia

> “ Black-cap sings sweetly, but rather inwardly: it is a songster of the first rate. ”
>
> GILBERT WHITE, Journal, May 19, 1770

▲ Awaiting a meal
Juvenile Lesser Whitethroats (*C. curruca*) are fed mainly on caterpillars, aphids, and flies by their parents. The young fledge 12–13 days after hatching.

ABYSSINIAN CATBIRD
Sylvia galinieri
This Ethiopian endemic has one of Africa’s most beautiful bird songs. The male sings a loud, ringing song, and the female harmonizes, interspersing a purring rattle.

GARDEN WARBLER
Sylvia borin
This very plain bird is a long-distance migrant from northern Europe and Central Asia to sub-Saharan Africa, reaching as far as South Africa.

DARTFORD WARBLER
Curruca undata
The Dartford Warbler nearly vanished from the UK, at the north of its range, with only 11 pairs left in 1963. This population has now increased to more than 3,000 pairs.

◄ Well camouflaged
The mottled brown wings and back of the Eurasian Treecreeper (*Certhia familiaris*) help it blend into its wooded habitat, protecting it from predators.

In flight, the **wallcreeper's** rounded crimson **wings** recall those of a **butterfly**

SULPHUR-BILLED NUTHATCH
Sitta oenochlamys
Restricted to the Philippines, this bird has a bright yellow bill and bare skin around its eyes.

WALLCREEPER
Tichodroma muraria
Native to the mountains of Eurasia, the Wallcreeper is the only living species in the family Tichodromidae.

AFRICAN SPOTTED CREEPER
Salpornis salvadori
Widespread in the open forests of sub-Saharan Africa, this species builds a cup nest on a branch.

Nuthatches, Wallcreeper, and Treecreepers

Sittidae, Tichodromidae, Certhiidae

Order Passeriformes

Families 3

Species More than 40

Size range 4–8 in (10–20 cm) long

Distribution North America, Eurasia, Africa

Thriving on nuts, seeds, and insects, these active birds are typically found clinging to trees or rocks, where they search for food with a long, pointed bill.

These three families share similarities in the way they maneuver around their habitats, but when it comes to feeding techniques the nuthatch family is the odd one out. Members such as North America's Red-breasted Nuthatch (*Sitta canadensis*) forage for nuts, seeds, and insects, and then cache their finds in gaps in the tree bark. Later, they hammer away at the nuts with their robust bill—their name derives from the Middle English word "notehache," which means nuthacker. Treecreepers and the Wallcreeper (*Tichodroma muraria*) are rather more careful, daintily probing then plucking their favored insect and spider prey from crevices in branches and rock faces.

Like many climbing birds, treecreepers have stiffened tail feathers for stability and support. This feature is not shared by nuthatches, which instead have a strong hind toe with a long curved claw that allows them to hang from branches and walk both up and down tree trunks. Treecreepers are only able to work their way upward, and then must fly down to start again.

Most treecreepers and nuthatches are arboreal, but the unique wallcreeper is found in rocky mountainous habitats. It has been known to breed at altitudes of up to 19,700 ft (6,000 m) or more, but descends during the winter months.

◄ Acrobatic nuthatches
These White-breasted Nuthatches (*Sitta carolinensis*) demonstrate an array of climbing and perching skills in this 1832 illustration by Robert Havell after John James Audubon.

Swallows and Martins

Hirundinidae

Swallows and martins are known for their skill in the air, capturing insect prey on the wing in agile and tireless flickering, wheeling flight.

A diet of flying insects means that in cooler temperate regions swallows and martins can only survive in the warmer months. In winter, they migrate to warmer climates, and their yearly movements are much-celebrated signs of seasonal change. However, the annual arrival of migratory swallows and martins has also inspired some cautionary tales, including Aesop's fable *The Spendthrift and the Swallow*, in which a young man sells his cloak on seeing an early spring swallow—only for both man and bird to perish in a subsequent cold snap. The proverb "One swallow doesn't make a summer" gives the same warning.

The full truth of the huge distances covered by some swallows and martins was revealed in 1911–12, when wildlife enthusiast John Masefield attached bands to the legs of baby swallows on his property in the UK, and was contacted months later by a farmer in South Africa who had found one of the birds.

Like many birds that range over a wide area for food, swallows and martins tend not to be strongly territorial. They often nest in colonies on a cliff or high ledge, but increasingly on

The deeply forked tail stands out in flight

▲ **Shape of freedom**
The flying swallow is a popular motif, adorning everything from this Italian silk fan (c.1900) to the tattooed arms of sailors.

◄ **High-rise block**
American Cliff Swallows (*Petrochelidon pyrrhonota*) use mouthfuls of mud to build nests on high walls, often attaching to their next-door neighbor.

Thumbelina
In Hans Christian Andersen's fairy tale of 1835, a friendly swallow saves Thumbelina from an unwanted marriage to a mole.

Order Passeriformes

Families 1

Species More than 80

Size range 4–9½ in (10–24 cm) long

Distribution Worldwide (except Antarctica)

SLEEPING WITH THE FISH

Early European naturalists such as Aristotle wondered whether barn swallows and other migratory birds overwintered underwater in ponds, like frogs, before observations showed that they flew south in winter. Here, a pair of Swedish fishermen haul in a net full of swallows from a lake.

WOODBLOCK PRINT, C.1555

Swallows have **relatively long eyeballs,** which help them **hone in** on tiny prey

houses or other buildings. As familiar town and village birds, they are prominent in stories such as the 200-year-old Korean folk tale of brothers Heungbu and Nolbu. Heungbu was rewarded for rescuing a baby swallow, and the bird brought him seeds that grew into treasure-bearing fruit. However, another swallow also delivered seeds to the cruel and jealous older brother Nolbu, and the gourds grown from these seeds brought forth only destruction. Irish author Oscar Wilde's story *The Happy Prince* (1888) tells of a golden statue who teams up with a swallow to save a town from poverty and famine, the swallow delivering the statue's valuable gems and gold-leaf to the people.

Close quarters

Swallows and martins usually nest in single-species colonies, but several species may feed together. When doing so, they show a particular behavior called "niche separation," in which each species feeds in a separate height zone to help them coexist. However, living in close proximity does mean that mating competition can be rife in a colony. For example, the rate of extra-pair paternity (proportion of chicks in a nest that are not the offspring of the male providing food) can exceed 50 percent in White-rumped Swallows (*Tachycineta leucorrhoa*), Violet-green Swallows (*T. thalassina*), and Tree Swallows (*T. bicolor*).

Mighty martin
The Purple Martin (*Progne subis*) is a powerful bird that migrates from North America to South America for winter.

BARN SWALLOW
Hirundo rustica
Widespread and well known, this species breeds across the Northern Hemisphere and winters across the Southern.

GREATER STRIPED SWALLOW
Cecropis cucullata
This large swallow breeds in Southern Africa and migrates to winter in Tanzania and adjacent countries.

SAND MARTIN
Riparia riparia
This small martin digs its nesting burrows in steep sandy or gravel banks. Each tunnel may be 3 ft (1 m) long or more.

▲ **On the alert**
The Long-billed Thrasher (*Toxostoma longirostre*) perches in shrubs and trees to eat berries, but also forages on the ground.

▲ **Snake drama**
Robert Havell's 1827–30 painting shows northern mockingbirds vigorously defending their nest from a rattlesnake.

Mockingbirds and Thrashers

Mimidae

These graceful, noisy, long-tailed songbirds include several species that make up a conspicuous part of the garden bird life in North America.

The Northern Mockingbird (*Mimus polyglottos*) is much loved in the US for its beautiful song and talent for mimicry. Indeed, the third US president, Thomas Jefferson, was so fond of the species that he kept several as pets in the late 18th century to enjoy the song they provided. The northern mockingbird is also probably the species referenced in the book title *To Kill a Mockingbird*, by Harper Lee (1960), in which lawyer Atticus Finch uses the titular term as an analogy for the destruction of innocence.

Make some noise

Other well-known mockingbirds and thrashers in North America include the Gray Catbird (*Dumetella carolinensis*), with its secretive habits and loud mewing call. It has been heard mimicking other birds, frogs, and mechanical sounds, and strings together random phrases to make a lengthy song. The Brown Thrasher (*Toxostoma rufum*) draws attention to itself by rummaging noisily in leaf litter. The family is also well-represented in the Caribbean. And on the Galápagos Islands, a single mockingbird species that colonized sometime between 0.6 and 5.5 million years ago has since diverged into four distinct species. With their fearless and aggressive behavior, these species are quickly noticed by island visitors. In the 19th century, Charles Darwin's observations of these birds, along with the Galápagos finches, inspired him to develop the theory of evolution by natural selection.

Order Passeriformes

Families 1

Species More than 30

Size range 8–13 in (20–33 cm) long

Distribution North and South America

Survival strategy
A group of fluffed-up tree swallows snuggle together for warmth on a branch overlooking the Yukon River, northwestern Canada, in mid-May. Having flown thousands of miles on their spring migration from Central America, they had been surprised by a sudden snowstorm. Fortunately, the group huddle helped them survive the sub-zero temperatures and they lived to fly another day.

SOMBER GREENBUL
(*Andropadus importunus*)

Many of the drabber bulbuls are known as "greenbuls," but this species stands out for its particularly striking white eyes. It occurs across coastal eastern and southern Africa.

MOUNTAIN BULBUL
(*Ixos mcclellandii*)

This forest-dwelling bulbul, with its bushy crest and silvery "beard," occurs mostly above 2,600 ft (800 m). It is represented across a broad swathe of Southeast Asia by nine distinct subspecies.

RED-TAILED BRISTLEBILL
(*Bleda syndactylus*)

This bristlebill is found in dense forest in West and Central Africa. It predominantly forages for insect prey on the ground.

Indian nightingale
This engraving by James Forbes, published in his *Illustrations to Oriental Memoirs* (1835), shows a red-vented bulbul, a sweet-singing species still often kept as a pet "nightingale" in India.

The name "bulbul" is synonymous with "singing bird" in some Asian cultures

Bulbuls

Pycnonotidae

Order Passeriformes

Families 1

Species More than 160

Size range 5–11½ in (13–29 cm) long

Distribution Africa, Middle East, Asia

This large family includes a few conspicuous species whose sweet melodious songs are a familiar sound across the towns and gardens of many African and Asian countries.

Although not colorful, most bulbuls are lively and vocal, and some have very beautiful songs.
A Persian version of the folk tale *The Three Golden Children* features the *Bulbul-i-hazár-dástán* (bird of a thousand tales), which always sings the truth. Some bulbuls are popular pets, and the Red-vented (*Pycnonotus cafer*) and Red-whiskered Bulbuls (*P. jocosus*) have established invasive populations in some parts of the world from the offspring of escapee caged birds. Many others are rare and shy forest species, which are difficult to observe and even trickier to identify. Bulbuls are key dispersers of seeds. In the Middle East, the distribution of the mistletoe species *Plicosepalus acaciae*, which grows on acacias, is linked to the movements of the Yellow-vented Bulbul (*P. goiavier*).

Intelligent and adaptable, some bulbuls have found ways to deal with cuckoo parasitism. In China, the Brown-breasted Bulbul (*P. xanthorrhous*) avoids habitats shared with the Himalayan Cuckoo (*Cuculus saturatus*), while the Collared Finchbill (*Spizixos semitorques*) can spot and eject a cuckoo egg from its nest.

◄ Talking heads
Red-whiskered bulbuls raise their long crests to signal agitation or alarm. Native to Asia, this species is now naturalized in Florida, Hawaii, and Mauritius.

▼ National treasure
The Common Bulbul (*P. barbatus*) is the national bird of Liberia, where it is also known as the pepper bird due to its fondness for eating cultivated peppers.

Stamp from c.1953

Starlings

Sturnidae

All sheen, strut, squawk, and squabble, starlings are hard to ignore—especially as many of them have a long history of sharing space with humans.

▲ Myna detail
Two Common Hill Mynas (*Gracula religiosa*) rest in a tree in this inset of a 16th-century manuscript about the birds of Hindustan (a name for the Indian subcontinent).

The dusk murmurations of Common Starlings (*Sturnus vulgaris*), when huge flocks gather and swirl around in synchrony before going to roost together, is one of nature's most inspiring sights. For the ancient Romans, the patterns that these common starlings formed in the air contained prophecies, while for today's ornithologists, the meaning behind the movements is still somewhat mysterious, although the dazzling and unpredictable mass movement may help them confuse predators as well as reinforce social bonds. This species is native to Eurasia but has been widely introduced elsewhere, including North America, South Africa, and Australia, where it is sometimes viewed as an invasive species. It is one of the best-known members of this lively, noisy, and often colorful family.

Order Passeriformes

Families 1

Species More than 120

Size range 14–20 in (35–51 cm)

Distribution Europe, Africa, Asia, Australasia, Oceania

MOZART'S STARLING

In 1784, Wolfgang Amadeus Mozart purchased a common starling from a pet shop in Vienna. He taught the bird to whistle certain musical phrases, including the beginning of the third movement of his own piano concerto No. 17 in G major, which it replicated almost exactly. When the starling died three years later, he held an elaborate funeral for it, with veiled mourners singing a requiem.

STARLING'S RENDITION TRANSCRIBED BY MOZART

PIANO CONCERTO NO. 17 IN G MAJOR

VIOLET-BACKED STARLING
Cinnyricinclus leucogaster

This species lives in wooded parts of sub-Saharan Africa and rarely forages on the ground. Only the male has vivid coloration, with the female having brown streaked plumage.

> " Here rests a bird called Starling,
> A foolish little Darling. "
>
> WOLFGANG AMADEUS MOZART, "Funeral Poem," 1787

Masters of mimicry

Most starlings and mynas forage on the ground, searching and probing for invertebrate prey. They tend to be social when not nesting, and many species have a talent for mimicry. They incorporate these sounds into their songs, as a more varied song is attractive to potential mates. Pet starlings of various species will imitate speech and other sounds readily. Hill mynas (*Gracula* spp.), often kept as pets, are well-known for being capable of accurately imitating human speech, although there is as yet no evidence that they show quite the same level of understanding of their own utterances that has been documented for some talking parrots. Nevertheless, talking mynas appear in various legends and folktales, typically deceiving a figure of authority. In an old Chinese folktale, Liu Shan's pet myna not only helped his owner avoid debtor's prison, but also thoroughly humiliated the court magistrate by taking on the identity of one of the local gods.

The cage-bird trade has imperilled several of these species, including the various glossy starlings of southern Africa with their dazzling iridescent plumage, and the pure white Bali Myna (*Leucopsar rothschildi*), a species now reduced to fewer than 50 birds in the wild. Conversely, escapee pet Common Mynas (*Acridotheres tristis*), native to southern Asia, have established themselves as an invasive species in many regions around the world. The rhabdornises (*Rhabdornis* spp.), found in the Philippines, are probably the most unusual members of the family. These forest birds have camouflaged plumage and climb tree trunks, probing the bark for food, like treecreepers (a family that is not present in the Philippines).

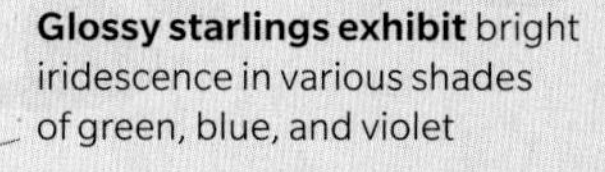

Glossy starlings exhibit bright iridescence in various shades of green, blue, and violet

◄ Riot of color
A common species in eastern Africa, the Superb Starling (*Lamprotornis superbus*) can often be found around towns and villages, but also in large flocks in the open savanna.

BALI MYNA
Leucopsar rothschildi

This beautiful bird is prized as a cage bird, which led to its catastrophic decline on Bali, the only place it naturally occurs. It often nests in an abandoned woodpecker hole.

ROSY STARLING
Pastor roseus

With bright plumage and a long crest, this striking starling often travels in vast flocks. It breeds in Central Asia and generally winters farther south, but can often be seen far from its usual range.

A model murmuration
Thousands of common starlings swirl in synchrony above Lough Ennell, Ireland, as they prepare to roost in the reedbeds along its shore. The high speed at which a flock turns makes it looks as though the birds share a single mind. However, in the early 2000s, Nobel laureate Giorgio Parisi used 3-D modeling to show that the effect is achieved by each starling responding quickly to changes of direction by those few birds flying right next to it.

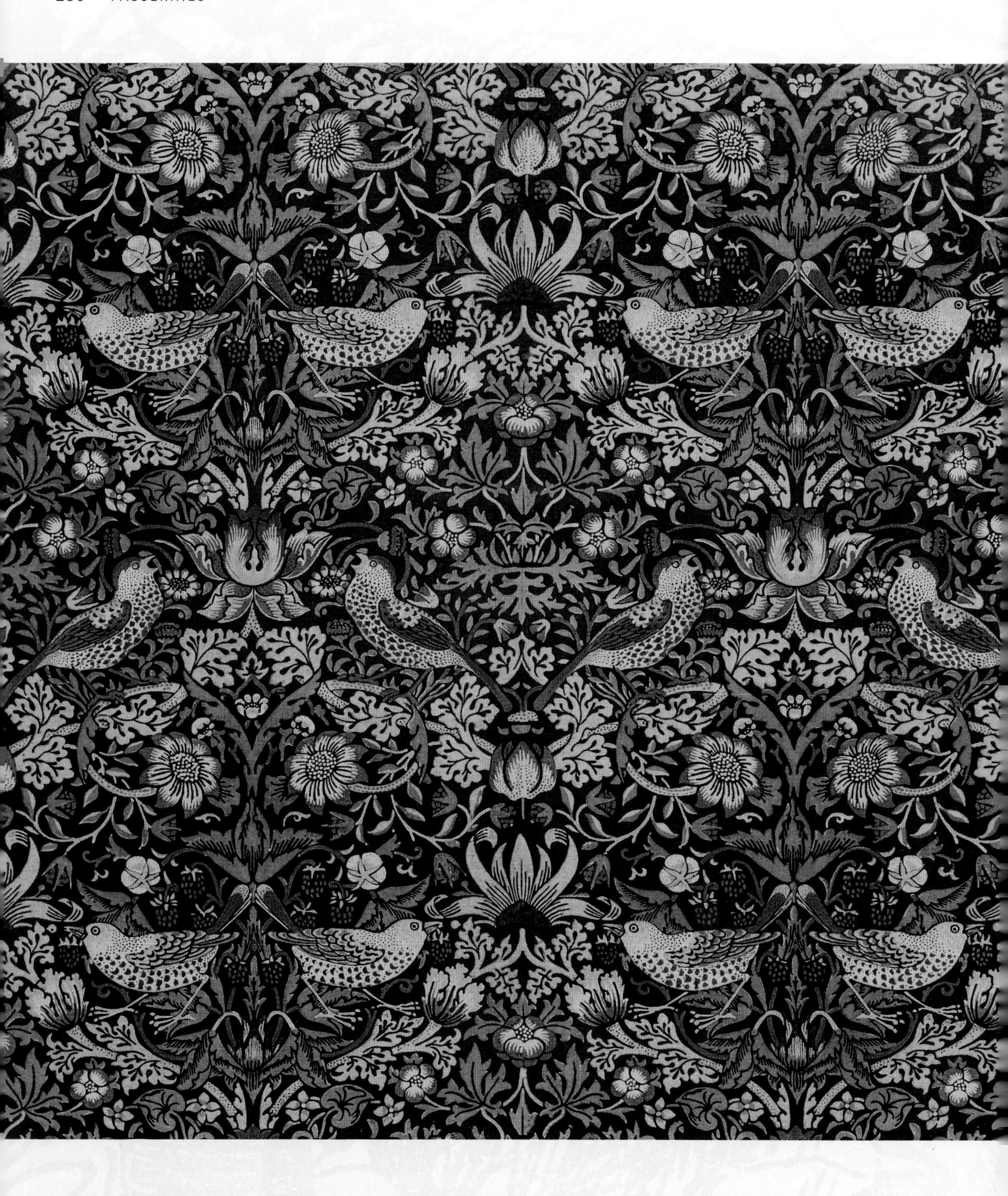

Order Passeriformes

Families 1

Species More than 170

Size range 5–13 in (12–32 cm) long

Distribution Worldwide except Antarctica

◄ **Strawberry thieves**
This dazzling tapestry was made in 1883 by British artist William Morris. It was inspired by the unwelcome sight of thrushes feeding on strawberries in his garden.

▲ **Filling the tank**
With a long migration ahead, an American Robin (*T. migratorius*) must consume extra food (and reduce its activity levels) to build up a surplus of body fat to fuel its autumn journey.

Thrushes

Turdidae

These tuneful songbirds are found in all kinds of habitats, including gardens. Many are notable long-distance migrants, while others are shy but colorful forest-dwellers.

Despite their **scientific name,** not all **American robins migrate,** staying on if food is **plentiful**

The typical thrush is a sleek, rounded bird that moves on the ground with a relaxed, bouncy hopping gait. It eats a mixture of fruits and invertebrate animals, and its plumage is probably mostly brown, perhaps with some red or blue-gray accents and some speckling—although there are exceptions, such as the stunningly bright-blue male Grandala (*Grandala coelicolor*). It is also likely to be a fine songster, with a repertoire of mellow and full-throated fluting phrases. Thrushes are larger than most of the passerines that share their (usually wooded) habitats. In general, they are territorial birds, although some will form large flocks in winter.

Poetic inspiration

The well-named Song Thrush (*Turdus philomelos*) of Eurasia and North Africa is the likely subject of Thomas Hardy's poem "The Darkling Thrush" written in 1900. Through its joyous song, the "aged thrush, frail, gaunt, and small" revives the poet's flagging spirits on a desolate winter's day. This bird produces a song noted for its repeated phrases as well as its beauty. Another English poet, William Ernest Henley, was similarly moved by a different species of thrush, the Common Blackbird (*Turdus merula*), whose song "is all of the joy of life."

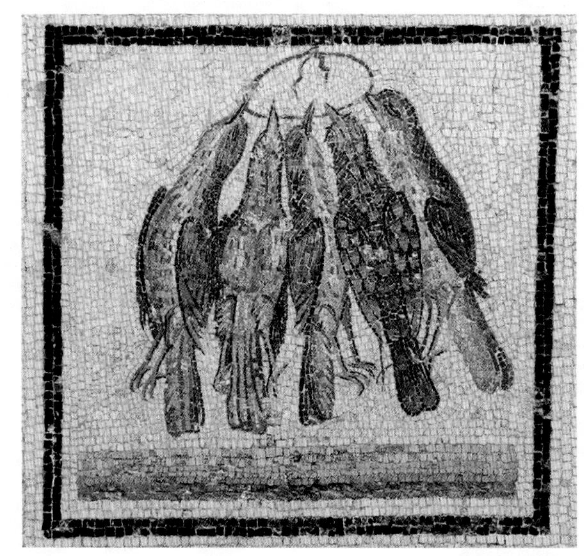

◄ Five for the pot
This Roman mosaic, from Thysdrus (modern-day El Djem, in Tunisia), shows thrushes that have been trapped for eating. Thrushes and other songbirds, served plucked but whole, regularly featured at ancient Roman feasts and festivals.

Hitching a ride

In North America, the nightingale thrushes (*Catharus* spp.) are represented by several rather small, speckled woodland-loving thrushes with notably lovely songs. The Hermit Thrush (*C. guttatus*) in particular is a celebrated songster, imbuing its notes with a dreamy beauty that has inspired several folk tales. One comes from the First Nations Oneida people, in which the bird learned its song after sneaking onto the back of an unsuspecting eagle in a high-flying contest. The winning bird would be awarded the sweetest song, and thanks to its tactics the hermit thrush was the victor. Its free ride carried it to a spirit realm, filled with beautiful music. The thrush became ashamed, though, of having cheated its way to that higher place, and also feared reprisals from the eagle. As a result, the hermit thrush will only ever sing its heavenly song when it is safely hidden away.

> " Morning has broken, like the first morning
> Blackbird has spoken, like the first bird. "
>
> ELEANOR FARJEON, "Morning has broken," 1931

The best-loved thrush in North America is probably the American Robin (*T. migratorius*), which is also the continent's largest thrush species. One of its claims to fame is an appearance (in puppet form) in the 1964 musical film *Mary Poppins*—which is set in London, UK, on the other side of the Atlantic from its native range. American robins, like their cousins the bluebirds (*Sialia* spp.), are often long-distance travelers within the Americas, with the northernmost breeding populations traveling hundreds of miles to reach wintering grounds farther south.

In recent years, the three bluebird species have faced habitat loss due to urban sprawl, as well as competition from invasive bird species and predation by domestic cats. Naturalists have successfully encouraged people to support bluebirds by building and maintaining bluebird boxes, where many now nest.

► Battle for survival
Both the fieldfare (top) and the Mistle Thrush (*T. viscivorus*, bottom) will defend a food source in icy winter weather. The birds are similar in size, so sometimes disputes may only be settled by a physical fight.

The blue color is produced by light being refracted by tiny pockets of air in dark feathers rather than pigmentation

▲ Bright and beautiful
The Eastern Bluebird (*S. sialis*) is a frequent garden visitor, and it will often breed in specially built nest boxes.

BASSIAN THRUSH
Zoothera lunulata
This large Australian thrush lives in damp forested areas, feeding on invertebrates it finds in the leaf litter. Its mottled-brown plumage provides excellent camouflage.

GRANDALA
Grandala coelicolor
A high-altitude forest dweller, this dazzling Himalayan thrush feeds on insects and fruits it finds on trees. The female is brownish with a white wing stripe.

PIED THRUSH
Geokichla wardii
This strikingly marked thrush lives on the Indian subcontinent. Adult males have black-and-white plumage; females have a similar pattern but in shades of brown.

Swainson's Thrush (*C. ustulatus*) is another North American species that travels south for the winter. Like many other migratory birds, it travels by night when the danger from aerial predators is reduced—a starry sky may also help with navigation. Flying by night and refueling by day leads to serious sleep deprivation, and a lengthy sleep in daylight hours, when hawks and falcons are hunting, would be risky. Studies on the brains of Swainson's thrushes have shown that the birds cope by engaging in hundreds of "microsleeps" throughout the day, each lasting only a few seconds.

In Eurasia, the Redwing (*T. iliacus*) is a noted migratory thrush, heading south or west from Scandinavia, Iceland, and Russia in fall in search of bountiful berries. The eerie rushing wings of flocks on the move make a distinctive sound that fishermen on the North Sea called "the herring spear"—the sound overhead might coincide with big hauls of herring, though it also struck fear into the men's hearts. Another Eurasian migrant, the Fieldfare (*T. pilaris*), is noted for its extreme self-defense tactics when danger threatens its nest. Several pairs nest in close proximity, and if one bird raises the alarm, all rise up and spatter the approaching would-be predator with droppings. This bombardment scares away most and can even render an intruding bird flightless.

Tropical variety

Thrushes are also present in the tropics, with the Island Thrush (*T. poliocephalus*) being present on many islands in Malaysia and Indonesia. Its various populations have been separate for so long that many have distinct traits and are now classed as subspecies. In all, the island thrush has nearly 50 recognized subspecies, which vary from black through various combinations of gray and reddish-brown, with some having pure white heads. Other Asian thrushes include the strikingly

▲ Mountain dweller
This stamp from St. Vincent depicts a Rufous-throated Solitaire (*Myadestes genibarbis*), a thrush of high-altitude forest on several Caribbean islands.

More than **50** bird species **collect shell pieces** from song thrush anvils as a source of **calcium**

SMASH AND GRAB

Song thrushes have learned how to break snails' shells, by gripping them on the rim and using a fast head-flick to strike them forcefully against a suitable stone. Individual birds have favorite stones that they use for this purpose—a well-used "thrush's anvil" is easily identified by the bits of broken shell that surround it.

SONG THRUSH AT ITS ANVIL

◄ Look both ways
This 1835 painting from John James Audubon's *Birds of America* depicts a pair of Varied Thrushes (*Ixoreus naevius*), perched below a superficially thrushlike but unrelated Sage Thrasher (*Oreoscoptes montanus*).

colored cochoas (*Cochoa* spp.), with bright green, blue, or purple plumage, and the highly gregarious, electric-blue grandala of the Himalaya.

Mistaken identity

The family is also well represented in Africa, with some groups, such as rufous thrushes (*Stizorhina* spp.), found only there. A Kenyan folk tale tells that Ngai, a supreme deity, gave the thrush the daily task of awakening the world's human inhabitants, and this is why thrushes begin to sing earlier in the morning than other birds. The species, in this case, may be the widespread African Thrush (*T. pelios*) or a related species, but could also be the Spotted Palm Thrush (*Cichladusa guttata*), which is not a true thrush but a chat (family Muscicapidae). Chats and thrushes are generally rather similar in appearance and habits and were once grouped in the same family, but examination of their DNA has revealed that this was an error. It persists in the common names of several species, including the rock thrushes (*Monticola* spp.) and the palm thrushes, which are chats, and the African Boulder Chat (*Pinarornis plumosus*), which is a thrush!

Special commission
In this natural history study by Indian artist Shaikh Zain al-Din, an Orange-headed Thrush (*Geokichla citrina*) perches in a purple orchid tree above a lesser death's head hawkmoth. Commissioned in 1778 by Lady Mary Impey in Calcutta (present-day Kolkata), it was one of 326 paintings of native birds and other animals in her collection.

Order Passeriformes

Families 1

Species More than 350

Size range
4–12 in (10–30 cm) long

Distribution
N. America, Europe, Africa, and Asia

➤ Vision in red
The Common Redstart (*Phoenicurus phoenicurus*) is a summer visitor to Eurasia, wintering in Africa. It is named for its red tail—"start" comes from the Old English word "steort," which means tail.

Chats and Flycatchers

Muscicapidae

This large and varied family of small insectivores includes one of Europe's most easily recognized and much-loved songbirds, the cheeky robin.

Chats and flycatchers may be found in habitats as diverse as Arctic tundra, humid tropical rainforests, arid semideserts, and high mountain slopes. They have made the most of these varied regions by specializing in the exploitation of a truly abundant source of food. The family name, Muscicapidae, originates from the Latin words "musca" (a fly) and "capere" (to catch). As this implies, birds of this family specialize in taking insect prey. Many species sally out from a prominent perch to catch their prey while in flight, audibly snapping their bill shut on the hapless insect, before returning to the perch. Others, however, will hop along the ground and pounce on their prey, or glean it from branches and leaves.

Feathered conundrums

Chats and flycatchers form one of the largest of families within the passerines. Scientists have long argued about their classification (for example, chats used to be grouped in the same family as thrushes). This is

Despite the family's **success** in the Old World, only **two species** also **breed in the New World**

➤ Tail of two halves
Spotted Forktails (*Enicurus maculatus*), as depicted in this 1850 illustration by John Gould, feed around the fast-flowing streams of South and East Asia.

primarily because members of this family have no single defining feature that sets them apart from other families of birds.

With a huge range of plumages, forms, behaviors, and habitat preferences on show, this is one of the most diverse of all groups of birds. Some members of this family, such as the scrub robins (*Cercotrichas* spp.), spend most of their lives on or close to the ground, frequenting dense patches of thicket. Many others are predominantly arboreal, found in the midstory or canopy of trees and only rarely venturing closer to the ground.

Some genera, such as the *Ficedula* flycatchers, are strongly sexually dimorphic—in other words, the males and females have very different plumages—but in others, such as the closely related *Muscicapa* flycatchers, both sexes look the same. And while many species have unmistakable vibrant or even iridescent plumages, others are much more subdued with subtle tones of earth-brown or gray.

A good number of birds in this family are famous for their beautiful vocalizations, while others are largely silent or utter little more than

▲ Fairy standoff
The European robin features heavily in British folklore. In *Cock Robin Defending his Nest* (c.1850), John Anster Fitzgerald presents the bird alongside a number of fairy figures.

insignificant squeaks or clicks. Many are resident, never leaving their small territories during their entire lives, while others may migrate many thousands of miles between breeding and wintering grounds.

Even their nesting habits vary. Most build cup-shaped nests, made of grass, moss, and twigs, but the placement of these differs considerably. Some are cavity nesters, taking advantage of small openings in trees, stumps, and rock ledges, while others will exploit the old unused nests of other birds, and some even nest on the ground.

Among this family of contrasts, there are many exceptional birds. The Northern Wheatear (*Oenanthe oenanthe*), which breeds widely across the Northern Hemisphere, has the longest migration of any songbird. Some Alaskan populations travel as far as 9,000 miles (15,000 km) to reach their sub-Saharan African wintering grounds—an incredible feat for a bird that weighs only 9⁄10 oz (25 g).

Meanwhile, the Common Nightingale (*Luscinia megarhynchos*), a summer visitor to Europe and Central Asia that spends the winter in Africa, has long been celebrated for its rich and dynamic song, featuring in poetry, music, and literature for many thousands of years. Ancient poets such as Homer wrote about it, as did, more recently, Shakespeare and Keats. This species was also the subject of the first-ever commercially available recording of an animal, made by Karl Reich in 1910 in his aviary in Bremen, Germany.

The gardener's friend

Although many species in this family are shy, elusive birds that are difficult to observe in their heavily vegetated habitats, plenty of others are plucky and often confident in the presence of humans—few are more so than the European Robin (*Erithacus rubecula*), one of Europe's most recognizable birds. It is well known for being a reliable presence in many gardens across the continent, where it will happily cohabit with people and often builds its nest in seemingly unlikely places in or around homes. The bird has also been a prominent symbol of Christmas since the Victorian era, regularly featuring on decorations and cards. This familiarity undoubtedly contributed to it being voted the UK's national bird in 2015.

Under pressure

A range of human-related pressures has led to more than 15 percent of all members of this family being threatened with extinction. The most significant of these pressures, affecting some 60 species, is habitat loss. Many of them rely heavily on mature and well-established vegetation, which deforestation and clearance imperils in many parts of the globe. An example is the case of the Rufous-headed Robin (*Larvivora ruficeps*), a beautiful slate-gray bird with a stunning orange head, white throat, and powerful song, which breeds in a tiny area of Sichuan, China. Habitat loss through clearance and the building of dams is thought to be driving its already small population toward extinction.

For other species, such as the White-rumped Shama (*Copsychus malabaricus*), their stunning vocalizations are also their curse. Relentless illegal trapping in the wild for the cagebird trade, simply so people can enjoy their songs, poses a serious threat to many populations of once-common birds. For some of the rarer species, some found on a single island, introduced predators such as rats have also taken their toll.

VERDITER FLYCATCHER
Eumyias thalassinus
Named for the vivid blue plumage of the adult male, this species is found in the Himalaya and Southeast Asia, Borneo, and Sumatra. Females are a paler blue.

COMMON NIGHTINGALE
Luscinia megarhynchos
This species is well known for its beautiful song. During the day, the male sings to communicate with rival males, and at night to attract females.

> "Nothing in the world is quite as adorably lovely as a robin when he shows off . . ."
>
> FRANCES HODGSON BURNETT, *The Secret Garden*, 1911

A glossy blue-black head and back also identifies this bird as male

The male has the long graduated tail, whereas the female's is shorter and less ornate

The distinctive white rump gives the species its name

◄ Always on song
The white-rumped shama is well known for its rich song repertoire in India and Southeast Asia. The male is brightly colored, while the female is less colorful.

Male nightingales have a repertoire of up to **200** different **song types**

Order Passeriformes

Families 2

Species More than 90

Size rang 3½–9 in (9–23 cm) long

Distribution North and South America, Eurasia, North Africa

➤ **Nest inspection**
This 1833 engraving shows Sedge Wrens (*Cistothorus stellaris*) on a nest. Males build the exterior of multiple nests and females choose their preferred option.

BROWN DIPPER
Cinclus pallasii
This is the dipper of eastern Asia, from the Himalaya to Japan and eastern Russia. Like other dippers, it lives on fast-flowing streams, feeding on aquatic invertebrates.

WHITE-CAPPED DIPPER
Cinclus leucocephalus
This species feeds in rocky, fast-flowing streams. It wades in and swims but, unlike its Northern Hemisphere equivalents, it does not completely immerse itself underwater.

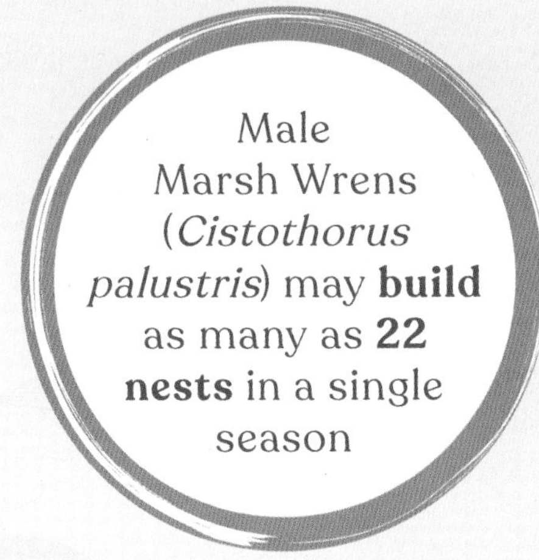

➤ Hunting by sound?
This Gray-breasted Wood Wren (*Henicorhina leucophrys*) is hunting for insects, possibly by listening for movement beneath the leaves, a technique that has been observed in other passerines.

Wrens and Dippers

Troglodytidae, Cinclidae

These two families contain plump, round-winged, short-tailed songbirds with loud voices. Wrens are mainly found low down in dense shrubbery, while dippers are aquatic.

The wrens of the Americas outnumber the rest of the world by 93 species to one, the outlier being the Eurasian Wren (*Troglodytes troglodytes*). This is a family of species with mainly brown plumage, often with bars, spots, and speckles. Their short wings enable them to maneuver within tight spaces in dense vegetation, where their narrow, decurved bill is ideal for picking insects from crevices. Their retiring habits demand loud voices for communication. Many wrens are renowned for their vehemence and extensive repertoire; in several species, males and females sing alternate phrases back and forth in what is known as an antiphonal duet. The song of the Musician Wren (*Cyphorhinus arada*) is delightfully bizarre.

Dippers are the world's only aquatic songbirds, native to fast-flowing streams and rivers. They search for aquatic prey, such as insect larvae and even small fish, often paddling with their head immersed. The northern species may submerge entirely and swim underwater for up to 30 seconds, using their wings as paddles. Like kingfishers, dippers have an extra eyelid, called a nictitating membrane, allowing them to see their prey underwater. They have dense plumage and a large preen gland with waterproofing oil. These extremely hardy birds will even feed under ice. The bulky, mossy nest is built in a crevice close to water and sometimes hidden behind a waterfall.

A short tail held cocked is typical of the family

➤ Loud for its size
Eurasian wrens are famous for the complexity and intensity of their songs—they seem too loud for such a small bird.

Standing guard
A male White-throated Dipper (*Cinclus cinclus*) keeps watch from his favorite vantage point on a rock in a waterfall in Levanger, Norway. A nest with his mate and their young lies behind the falls, and he is guarding them not only from predators, such as mink and rats, but also unmated dippers that might try to take over this prime piece of territory.

Flowerpeckers

Dicaeidae

Feasting on nectar and berries in Asia and Australasia, flowerpeckers are important for pollinating a wide range of trees and shrubs and dispersing their seeds.

Flowerpeckers do not really **peck flowers,** preferring **nectar** and **fruit**

▲ **Back-up plan**
This engraving by William Matthew Hart for John Gould's *Birds of Asia* (1850–83) depicts Scarlet-backed Flowerpeckers (*D. cruentatum*). During a food shortage, this species will go into a state of torpor (inactivity), behavior that is rare in passerines.

Flowerpeckers live in forested habitat from sea level up to an altitude of 12,000 ft (3,700 m). While some flowerpeckers are fairly plain, others have brightly colored males with distinctive patterns of gray, black, white, red, yellow, and orange. Often solitary or in pairs, they may gather in numbers at a fruiting tree, where several species might be encountered.

Flowerpeckers have a wide range of high call notes, described as chattering, buzzing, twittering, and trilling. Their songs include repeated ascending or descending phrases, and the Mistletoebird (*Dicaeum hirundinaceum*) is known to mimic other bird species. The first flowerpecker to be described, in 1747, was the Scarlet-backed Flowerpecker (*D. cruentatum*), then called the Black, White, and Red Indian Creeper. The most recently described species, the Spectacled Flowerpecker (*D. dayakorum*), was only discovered in 2009. The Cebu Flowerpecker (*D. quadricolor*) is the rarest—once thought to be extinct, it was rediscovered in 1992. Endemic to the Philippine island of Cebu, it is feared that fewer than 100 birds remain.

Flowerpeckers appear to be monogamous. The nest is usually a hanging pouch with a round or triangular entrance at the side. It is typically built by the female, but sometimes by both sexes, from mosses, spider webs, grasses, and small roots.

Mistletoe machines

Flowerpeckers feed on berries, pollen, and nectar, which forms an important part of their diet, and some eat seeds. They are active and agile, and their small size and short legs mean they can cling onto thin twigs to reach their food. A major disseminator of mistletoe seeds, flowerpeckers help many species of this parasitic plant to spread. Its seeds are sticky and will adhere to a branch when they have passed rapidly through the bird's digestive system. Flowerpeckers have a short, specialized gut that enables mistletoe berries to bypass the stomach and enter the gut directly so the seeds can be expelled intact and stick to a branch. Many species also eat insects and will hawk for them by flying out from a perch.

▲ Muted colors
Discovered in Borneo in 2009, the spectacled flowerpecker was formally described in 2019. Its scientific name, *dayakorum*, honors the local Dayak people, who conserve the forest in which it lives.

Order Passeriformes

Families 1

Species 50

Size range 2¾–5in (7–13 cm) long

Distribution Asia, Australasia

CRIMSON-BREASTED FLOWERPECKER
Prionochilus percussus
This colorful flowerpecker uses its stout bill to crush berries or puncture larger fruits so that it can "suck" or bite out pieces.

MISTLETOEBIRD
Dicaeum hirundinaceum
The Mistletoebird, the only flowerpecker in mainland Australia, can pass a mistletoe seed in as little as four minutes.

> " A lot of the times people assume that we've found most everything, especially with something as well studied as birds. "
>
> JACOB SAUCIER, *Audubon.org*, 2019

NECTAR GATHERER

Flowerpeckers that feed primarily on nectar have a specially shaped tongue that helps channel the sticky fluid into their throat. The shape varies among species. For example, the tongue of the Olive-capped Flowerpecker (*D. nigrilore*), shown here, has a flat base then curls into a tube, which splits into six fringed elements at the end to form a brushlike tip.

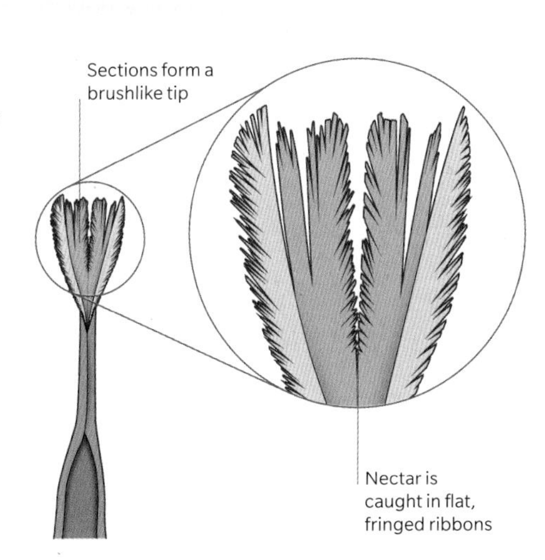

FLOWERPECKER TONGUE

Order Passeriformes

Families 1

Species More than 150

Size range 3–8½ in (8–22 cm) long

Distribution Africa, Asia, Australasia

Sunbirds and Spiderhunters

Nectariniidae

Found mainly in Africa and Asia, most members of this family rival hummingbirds in their vivid colors and iridescence, and they all share a love of nectar.

In many ways, sunbirds are the Old World equivalent of the hummingbirds of the Americas, but they evolved entirely separately from them. Like hummingbirds, they feed on nectar, but only a few species of sunbirds can hover in front of blooms while sipping; most sunbirds and all spiderhunters perch to feed. They are also highly adapted to their diet. They have a long, decurved, sharply pointed bill and long tongue capable of extending out beyond it. The tongue splits into two at the tip, where the edges are frayed, and draws in nectar by capillary action. Lower down, the tongue's edges curl in to form a tube and the nectar is sucked in. Sunbirds and spiderhunters have strong legs and sharp claws to enable them to cling tightly and get into position to feed from blooms while perched. At the base of the bill, two flaps cover the nostrils to keep pollen out.

▼ Fynbos feeder
A male Greater Double-collared Sunbird (*Cinnyris afer*) shimmers in the sun as it rests on top of a king protea flower.

Metallic blue band separates metallic green upper breast and bright red band

Underparts of the male are scarlet, but the female is gray all over

Pivotal pollinators

Sunbirds and spiderhunters are small and extremely active birds that move about very quickly. They have adapted to every flowery habitat in their ranges, from urban gardens to scrub above the treeline at 16,000 ft (4,900 m) altitude. Throughout their range, which also

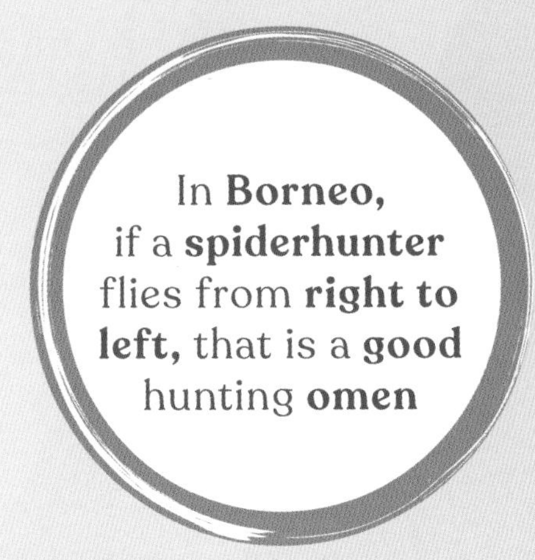

encompasses Australasia and Madagascar, they are important pollinators; it has been estimated that sunbirds provide this service to two percent of all South Africa's flora. This includes proteas found in the fynbos scrubland of the Cape. All species supplement their diet with a wide variety of invertebrates, including insects caught in midair, and many spiders plucked from cobwebs. So far, 70 species are known to take the latter including, not surprisingly, the spiderhunters.

Made for an audience

The glittering, showy plumage typical of the family is confined to males, with females having duller coloration. However, in some species the sexes are alike, with more modest plumage. Spiderhunters fall into the latter category; they are relatively large in comparison to sunbirds and tend to have longer bills.

As a rule, males sing and display to hold territories, females are responsible for nest-building and incubation, while both feed the developing young. A few male sunbirds form leks (a communal area for courtship displays), and one or two species show cooperative breeding. Sunbirds build a globular nest of grass, bark, leaves, and lichens, suspended from a low branch. It has a side entrance and is often intentionally far from neat, often with extra vegetation trailing down to hide the structure from predators.

Spiderhunters share incubation tasks and their nests are remarkable. They construct cup nests, which are often concealed by being attached to the underside of large leaves or a palm frond, bound there by cobwebs, grass, and sometimes sticks.

▲ Vertical climber
A Streaked Spiderhunter (*Arachnothera magna*) grips onto the developing flowers of a wild banana palm.

BROWN-THROATED SUNBIRD
Anthreptes malacensis
Found across Southeast Asia in open forest and urban areas, the Brown-throated Sunbird is regularly parasitized by cuckoos.

PURPLE SUNBIRD
Cinnyris asiaticus
A familiar sight over much of the Indian subcontinent and Southeast Asia, this common species often visits and breeds in gardens.

LONG-BILLED SPIDERHUNTER
Arachnothera robusta
With the longest bill of the family, three times the length of the head, this Southeastern Asian species can probe deep into vegetation.

Weavers

Ploceidae

Weavers are renowned for their expertly knotted and woven grass nests. The male builds the domed nest, which has an opening at the bottom or the side, and uses it to attract a female.

A fresh, green nest is more likely to be chosen by a female weaver and the male may build several before finding a mate

➤ Busy weaver
The male Village Weaver (*Ploceus cucullatus*) builds a nest, then hangs from it, flapping his wings to attract the attention of a female. If successful, he will add a tunnel entrance and the female will line the nest.

The adult male has a dark face mask

The bird chatters loudly and vigorously flutters its wings, making the nest swing

Order Passeriformes

Families 1

Species More than 120

Size range 4⅓–11 in (11–28 cm) long

Distribution Sub-Saharan Africa, Indian Ocean Islands (including Madagascar), Asia

Most members of the weaver family are similar in size and shape to the Old World sparrows, to which they are related. Some species are found in Madagascar and southern Asia, but most populate the open country and forests of sub-Saharan Africa, where weaving material is readily available. Denizens of the forest are insectivorous, with slender, pointed bills, while weavers of the open country have the small, thick bills characteristic of seed-eaters.

Woven wonders

As the common name implies, most of the birds in this family construct their nests by weaving together vegetation such as grass, reeds, or palm leaves. Some nests are large and messy with others small and neat. A single vertical support is used by some species, while others form a bridge between two supports. Male weavers instinctively build such nests using intricate knots, improving with practice. Many weavers start with a suspended loop that is then expanded into a wall on one side, and an opening on the other, while others suspend a cup nest, then add a roof and a side entrance. The construction of an outer shell may require 300 grass strips and while a simple nest may only take a day to complete, others may take a week or two. Nests are often constructed suspended from a branch that hangs over water, and it is thought that this placement might be a form of deterrent or protection against predators. Snakes, such as the boomslang, will prey on chicks, and even African elephants have been seen eating weaver nests whole.

Extremely sociable

Most weavers are highly gregarious, with the Red-billed Quelea (*Quelea quelea*) forming massive flocks. The Sociable Weaver (*Philetairus socius*) lives in a huge colony in a single domed nest that can contain as many as 300 individuals, each having their own nest hole. It is built from the top down, starting with a roof. The nest may weigh as much as a ton and can be 13 ft (4 m) in depth—the largest nest of any bird, with some in use for up to 100 years. In addition to nests, it also contains chambers used as dormitories throughout the year. Pygmy Falcons (*Polihierax semitorquatus*) have been known to breed and roost in disused chambers.

▲ Jumping Jackson
An undated painting shows a male Jackson's Widowbird (*Euplectes jacksoni*) in dancing display, jumping with head thrown back, tail plumes arched forward, and outer feathers lowered.

RED FODY
Foudia madagascariensis
The Madagascan Red Fody was introduced to other Indian Ocean islands, where it has become a pest and competes with endangered fodies on Mauritius and Rodrigues.

THICK-BILLED WEAVER
Amblyospiza albifrons
Also called the grosbeak weaver, this species has a stout bill for cracking hard seeds. Its intricate nests are woven from fine strips of grass or sedge.

Red-billed quelea flocks may contain as many as 30 million birds

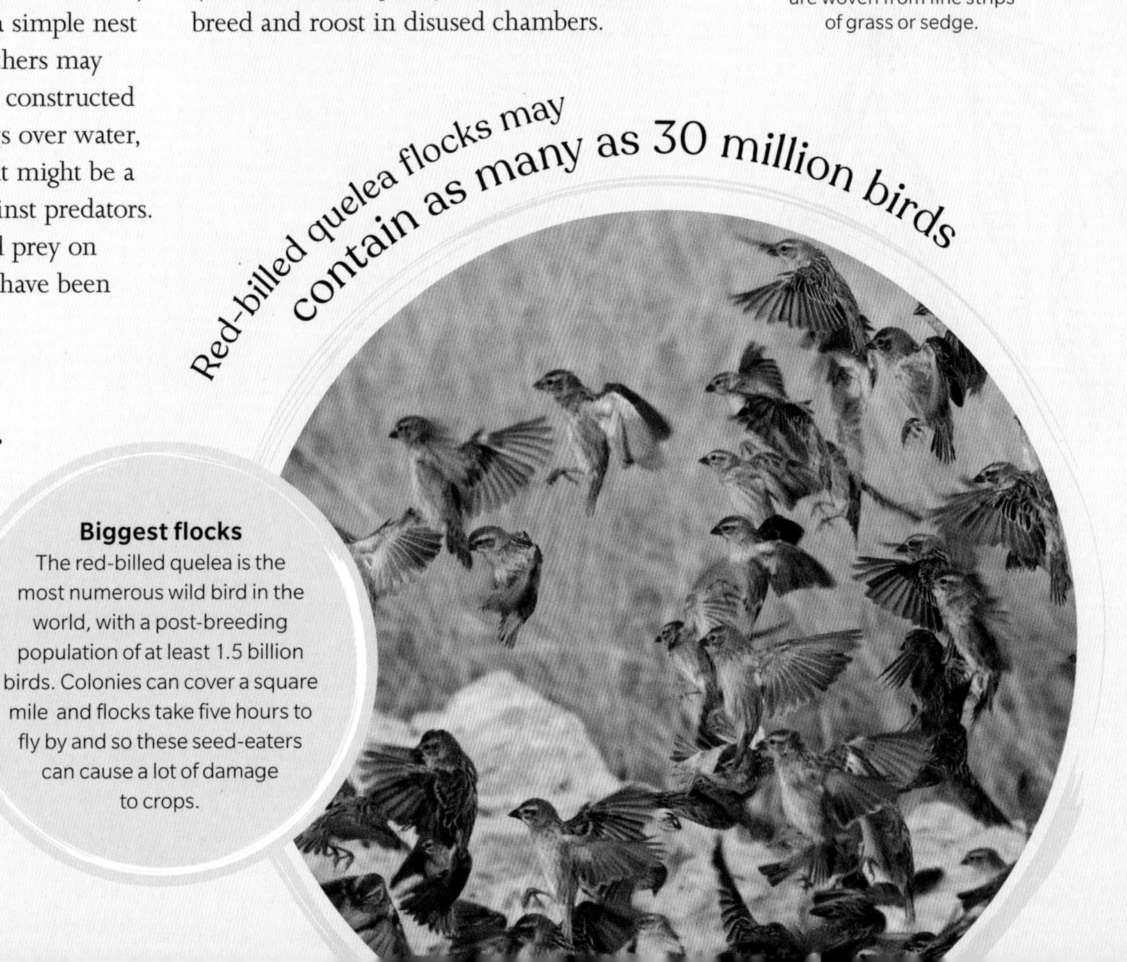

Biggest flocks
The red-billed quelea is the most numerous wild bird in the world, with a post-breeding population of at least 1.5 billion birds. Colonies can cover a square mile and flocks take five hours to fly by and so these seed-eaters can cause a lot of damage to crops.

▲ **Colorful faces**
Gouldian finches are found in three genetic variants. The most common (75 percent) have a black face, and the rest are mainly red-faced, with yellow-faced being very rare.

Waxbills

Estrildidae

This colorful family of finchlike birds is named for the red sealing-wax colored bills of some species. Being easy to keep, coupled with their bright plumage, makes them popular cagebirds.

Despite being named for the red color of their bill, this only applies to 30 species of waxbills—most have a black or silver bill. Waxbills are a very variable family, with some brown and gray species, others bright red or blue, and several strikingly multicolored or patterned with stripes or spots. Though most prefer warm, tropical habitats, some live in cooler places such as the highlands of New Guinea. A few species, such as the White-breasted Nigrita (*Nigrita fusconotus*), eat mainly insects and caterpillars, as well as fruits, but generally waxbills are seed eaters.

Populations of some waxbill species have waxed and waned throughout the centuries—and their diet of seeds has sometimes been a contributing factor. The Java Sparrow (*Padda oryzivora*) derives its Latin name from its tendency to raid rice paddies and has been persecuted by

Order Passeriformes

Families 1

Species More than 130

Size range 3–6⅔ in (8–17 cm) long

Distribution Africa, Asia, Australasia

COMMON WAXBILL
Estrilda astrild

The red breast patch of Common Waxbills acts as a signal of dominance. The brightest patches are a badge of status and are found on the highest-ranked birds.

RED-CHEEKED CORDON-BLEU
Uraeginthus bengalus

This attractive and colorful waxbill is widely kept in captivity. It is common in the wild from West to East Africa and was introduced into Hawaii.

farmers for this. Other species, such as the Red Avadavat (*Amandava amandava*), have seen feral populations boom outside of their native range due to escapees from cages. Although native to Asia, feral flocks can be found as far away as Spain and Hawaii. The Gouldian Finch (*Chloebia gouldiae*) saw a sharp decline in Australia in the 20th century due to trapping for the caged-bird trade and habitat loss from agriculture. Schemes to restore the finches' habitat—and the hollow trees they nest in—have brought the species back from the brink of extinction.

➤ Rice bird
The Java sparrow is now rare in the wild due to trapping, but it is widespread in captivity. It is bred in a variety of colors, such as the all-white form depicted next to a wild bird in this 1899 lithograph.

AUSTRALIAN ZEBRA FINCH
Taeniopygia castanotis

A common cagebird, this zebra finch is also a popular species for scientific research into birdsong, parental care, and other areas of study.

Old World Sparrows

Passeridae

These are the archetypal "little brown birds" of towns, gardens, and farmland. Cheery, chirruping, and squabbling, they forage on busy streets and often nest in or near buildings.

Sparrow imagery is linked to prosperity

▲Valuable icons
Eurasian Tree Sparrows (*P. montanus*) decorate this early 19th-century Japanese lacquer inrō (small purse or case) by Hara Yōyūsai.

The order Passeriformes, the "perching birds," takes its name from the Latin word for sparrow, *passer*, reflecting a deep familiarity between humans and sparrows, and a long shared history.

Old World sparrows flock together to feed on seeds and, especially when nesting, insects. As soon as humans began to cultivate arable crops, sparrows moved in to take their share of the harvest, which has sometimes made them unpopular. The House Sparrow (*Passer domesticus*) of Eurasia and northern Africa was introduced to North America in the 1850s, where it thrived. By the 1880s, some US states had launched campaigns to eradicate it. Michigan citizens received a small bounty for each sparrow they killed, and some enterprising people kept and bred the birds to collect extra rewards. Through the 20th century, continued efforts were made to control house sparrows, not only because they threatened crops but also because they displaced native species, such as bluebirds, tree swallows, and purple martins, from nest sites. Even so, the North American population is more than 80 million.

Old World sparrows live in open habitats, including savanna and desert, as well as around human habitation. The snowfinches form a distinct group, living in snowy mountainous habitats. Both sexes have camouflaged gray and white plumage, whereas the males of other Old World sparrow species generally have bold black, white, gray, and reddish markings. Most

Order Passeriformes

Families 1

Species More than 40

Size range 4½–7 in (11.5–18 cm) long

Distribution Europe, Asia, Africa, Australia, North America

With a gray head, females are less eye-catching than males

The bright chestnut plumage is a feature of both sexes

Raiding the orchard
The small bird stealing figs in this 11th-century BCE painting on an Egyptian tomb in Luxor is likely a male house sparrow of the local subspecies *niloticus*.

species live in flocks all year round. In the breeding season, males sing (often simply loud monotonous chirping) close to nest sites to attract females and keep other males away. However, pairs nest close together, defending only a small territory around themselves. Golden sparrows, in which males have bright yellow plumage, are especially gregarious, with thousands of nests in each colony.

Unconventional types

Most Old World sparrows are sedentary and form long-lasting colonies. Some, though, such as the African bush sparrows (*Gymnoris* spp.), are more nomadic, wandering long distances between the dry and rainy seasons. Intelligent and inquisitive, Old World sparrows readily exploit new foraging opportunities and unconventional nesting sites. Some are even known to weave their own small nests into the huge stick constructions of storks and other large birds.

> " Are not two sparrows sold for only a penny? But not one of them falls to the ground without your Father knowing it. "
>
> *Holy Bible*, Matthew 10:29

▼ Group activity
Across much of southern Africa, large flocks of Cape Sparrows (*P. melanurus*) gather to drink or search for food.

WHITE-WINGED SNOWFINCH
Montifringilla nivalis
The most widespread of the snowfinches, this large and handsomely patterned bird is at home on high mountains across Eurasia, breeding at altitudes of 4,900 ft (1,500 m) or more.

CINNAMON IBON
Hypocryptadius cinnamomeus
Recently reclassified after study of its DNA, the warblerlike Cinnamon Ibon is an atypical Old World sparrow that occurs in dense mountain forests on Mindanao, Philippines.

DESERT SPARROW
Passer simplex
The Desert Sparrow is regarded as a good luck omen by the Mozabite Berbers of Algeria, who leave holes in their adobe buildings to encourage it to nest there.

Finches

Fringillidae

A large and widespread family, the "true finches" are small songbirds with sweet voices and robust seed-cracking bills. Many are boldly patterned, with colorful plumages.

Order Passeriformes

Families 1

Species More than 230

Size range 4–10 in (9.5–25 cm) long

Distribution Worldwide except Antarctica

The term "true finch" differentiates members of the Fringillidae family from other superficially similar groups that are not closely related, such as the cardinals, tanagers, weavers, and waxbills, all of which have been mistakenly classed within the Fringillidae in the past. Genetic studies confirm that the finch family does, however, include the Hawaiian honeycreepers, which look quite different, having long slender bills that are used to probe flowers for nectar.

Specialist feeders

Most finches eat seeds through at least part of the year. Species such as the Hawfinch (*Coccothraustes coccothraustes*) and the grosbeaks have huge bills and a tremendous bite force—strong enough to crack a cherry stone, in the case of the hawfinch. Many, though, switch to an insect-based diet in spring and summer. This flexibility is why most finches are present all year round in temperate countries, adding color to winter gardens when birds that eat insects only have long since migrated to warmer regions. The crossbills are highly specialized feeders, eating pine seeds that they pry from cones using the elongated overlapping tips of their mandibles. Like other specialists, they travel to find food when necessary, and sometimes undertake large-scale nomadic movements, known as irruptions, to exploit a bountiful pine seed crop.

A bold wing pattern helps flying flocks stay connected visually

A slim bill is useful when feeding from spiky seedheads

STATUS SYMBOL

This hand-stitched cape from c.1821 belonged to Queen Kaahumanu, the favorite wife of King Kamehameha I of Hawaii. It features thousands of bright red and yellow feathers, collected from the ʻIʻiwi (*Drepanis coccinea*), also known as the scarlet honeycreeper, and the now-extinct ʻOʻo (*Moho nobilis*) and Hawaii Mamo (*D. pacifica*). Valuable capes like this denoted status and were only worn by those held in high esteem.

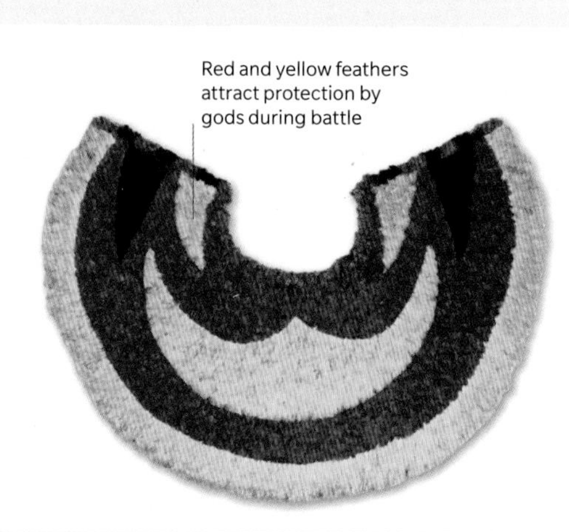

Red and yellow feathers attract protection by gods during battle

HAWAIIAN FEATHER CAPE

Madonna and Child
In this painting from c.1360–65 by Italian artist Lorenzo Veneziano, the European goldfinch is a living symbol of the Passion of Christ, its red face bearing the blood of the crucified Christ.

The goldfinch on the Madonna's hand may be interpreted as a symbol of the human soul being held "in the hand of God"

◄ American beauty
The striking American Goldfinch (*Spinus tristis*) is a regular visitor to gardens in winter. Its conical bill is well adapted for extracting and cracking small seeds.

The red crossbill (*Loxia curvirostra*) is one of several species said to have stained its feathers red while trying to remove Christ's crown of thorns during the Crucifixion. Another is the European Goldfinch (*Carduelis carduelis*), which has a bright red face. European goldfinches appear in around 500 religious works of art from the Renaissance period, although perhaps the most famous is Carel Fabritius's 1654 portrait *The Goldfinch*. The 2013 novel of the same name, by Donna Tartt, centers around a child who acquires this painting when the museum holding it is bombed by terrorists. European goldfinches are also popular as pets, but they are greatly outnumbered in captivity by the Atlantic Canary (*Serinus canaria*), which was domesticated in the 17th century. Today, more than 200 distinct breeds of canaries exist. Some are prized for the complex form of their song, while others have been modified by selective breeding to exhibit unusual plumage colors and textures, as well as exaggerated body shapes. The term "canary in the coal mine" stems from when miners in the UK used canaries (between 1911 and 1986) to detect emissions of carbon monoxide gas. If the caged canary collapsed, it was time to evacuate the mine. Canaries are also linked more generally with the idea of revealing truths. In film noir, a reluctant police informant might "sing like a canary" when bribed or threatened.

“ The young Bullfinches in their party colored Raiment . . . poize themselves like Wire dancers or tumblers ”

DOROTHY WORDSWORTH,
The Grasmere Journals, 1802

Lesser-known finches include the vividly colored euphonias and chlorophonias of Central and South America, which live mostly in dense rainforest, and the mountain finches and rosy finches, which inhabit high uplands in Asia and North America. In Costa Rica, the Golden-browed Chlorophonia (*Chlorophonia callophrys*) is known as the rualdo bird, and legend tells that it sacrificed its singing voice to a volcano in exchange for the life of a young girl.

▲ Breath of life
Used by miners, this canary cage includes an oxygen supply and pipe, so that a canary overcome by carbon monoxide might be saved at the eleventh hour.

Island life

The Eurasian Chaffinch (*Fringilla coelebs*) is a familiar bird to Europeans, but it also has numerous close relatives that are restricted to various Atlantic islands. The Tenerife Blue Chaffinch (*F. teydea*) is known for its striking smoky-blue plumage. Other endemic island species are threatened with extinction. The São Tomé Grosbeak (*Crithagra concolor*), for example, is classed as critically endangered, as are no fewer than 11 of the Hawaiian finch species (five may already be extinct). Threats to these and other native Hawaiian birds include the spread of avian malaria, habitat loss, and the impact of introduced predators, such as rats, cats, and the small Indian mongoose. Elsewhere, non-native finches have proved to be invasive. In New Zealand, European goldfinches, European Greenfinches (*Chloris chloris*), and Common Redpolls (*Acanthis flammea*) have all been introduced since Europeans arrived in 1769. All have thrived, to the detriment of native species.

◄ Handsome couple
This mid-19th century illustration by John Gould depicts a fine pair of Eurasian Bullfinches (*Pyrrhula pyrrhula*). They may raise up to three broods in a season.

► Mountain jewel
The golden-browed chlorophonia has a bold yellow eyebrow stripe and a vivid purplish blue cap. It feeds in trees and mostly eats fruit, including figs and mistletoe berries.

ATLANTIC CANARY
Serinus canaria
The ancestor of all pet canaries, the Atlantic Canary is native to the Canary Islands. Its natural plumage color is green, but captive variants can be white, gray, brown, variegated, or yellow.

JAPANESE GROSBEAK
Eophona personata
Weighing as much as 3½ oz (99 g), this handsome but rather shy bird is one of the world's largest finches. It can be seen in small flocks in forested landscapes in the far east of Asia.

HAWFINCH
Coccothraustes coccothraustes
The mighty bill of the Hawfinch can generate a biting force strong enough to crack almost any seed. The large head and neck contain impressive musculature to power this ability.

◄ **Unusual tones**
John Gould's 1837 engraving reveals the ashy blue and brick red coloration of the southern European Cretzschmar's Bunting (*E. caesia*).

Order Passeriformes

Families 1

Species More than 40

Size range 4¾–7½ in (12–19 cm) long

Distribution Europe, Asia, Africa

The yellow rump makes the bird easy to identify in flight

Unlike finches, buntings tend not to show bold wing markings

▲ **Asian wanderer**
A brightly colored bunting of Central Asia, the Red-headed Bunting (*E. bruniceps*) is a long-distance migrant that winters on the Indian subcontinent.

Old World Buntings

Emberizidae

These small finchlike songbirds are seed-eaters and close cousins to the New World sparrows. They prefer open, often rugged countryside, and some species sport boldly patterned plumage.

Over much of Eurasia and Africa, buntings are conspicuous birds across a variety of habitats, from marshlands to mountains. Males tend to deliver their rather monotonous and tuneless songs—that of the Corn Bunting (*Emberiza calandra*) sounds like a bunch of keys being shaken—from very visible elevated perches. They like to occupy a favorite spot for a long time, so can be easy to observe.

In most species, the male is more colorful and boldly patterned than the female, sporting striped face markings and patches of yellow, green, and rusty red. Females generally have excellent camouflage to disguise them while incubating their eggs, which they do in a beautifully constructed and well-hidden nest that is often on or close to the ground. The eggs are notable in some species for their complex patterning. The Yellowhammer (*E. citrinella*) used to be known as the "scribble lark" in England because of the random dark swirls and squiggles that decorate its eggshells.

After breeding, some buntings form large flocks, and those in more northerly areas will migrate south for the winter. One species, the Ortolan

Buntings **switch from seeds** to a diet of mostly insects when **feeding young**

Bunting (*E. hortulana*), has long been trapped by hunters as it travels in the Mediterranean. It is considered a great delicacy in France in particular, where diners traditionally covered their heads with a napkin, partly to capture the aromas and partly to conceal their identity from God. However, the ortolan has suffered severe population declines across Europe (down by 88 percent since 1980). Accordingly, hunting the bird and selling it in restaurants is now banned. A similar fate has befallen the Yellow-breasted Bunting (*E. aureola*) in eastern Asia, and this heavily hunted species is now critically endangered. The Cirl Bunting (*E. cirlus*) has fared better. It suffered heavy declines in the UK in the 20th century, but has now been successfully reintroduced to parts of its former range.

Lucky charms

As their common name suggests, House Buntings (*E. sahari*) are frequently encountered near domestic dwellings across southern Europe and northern Africa. In many towns and villages in Morocco, it is considered lucky to have house buntings around. And because the birds are much loved and encouraged, they can become quite fearless, regularly scrounging at diners' feet in outdoor venues. Some buntings, however, live in such remote regions that they rarely encounter people. The Tibetan Bunting (*E. koslowi*) is found only on the eastern Tibetan Plateau. Uniquely in the family, it constructs a roofed nest rather than an open one, presumably to give extra shelter in its chilly habitat. The Common Reed Bunting (*E. schoeniclus*) favors wetland habitats, nesting in dense reedbeds, but it often moves away to nearby farmland or even gardens in winter.

The name "bunting" has been applied to other groups of birds that are not part of the family Emberizidae but have some superficial similarities. Among these is the Snow Bunting (*Plectrophenax nivalis*), a resilient Arctic species from the family Calcariidae. In the Americas, five of the extraordinarily colorful songbirds that form the genus *Passerina* are called "buntings," but are actually members of the cardinal family.

HYBRIDIZATION

Although hybridization is generally rare in wild songbirds, it occurs regularly between the closely related pine bunting and yellowhammer, where their ranges overlap in northwestern Asia. The male hybrid (*E. citrinella* x *leucocephalos*) has a pine bunting-like head pattern, with bright coloration from its yellowhammer ancestry.

MALE HYBRID

YELLOWHAMMER
Emberiza citrinella
This European bunting, whose name incorporates the German word *ammer* (meaning "bunting"), is one of several UK songbirds introduced to New Zealand in the 1800s.

CRESTED BUNTING
Emberiza lathami
A large bunting with unusual coloration and a larklike crest, this species occurs in grassy areas in hilly parts of tropical and subtropical Southeast Asia.

> " Five eggs, pen-scribbled o'er with ink their shells . . . They are the yellowhammer's "
>
> JOHN CLARE, "The Yellowhammer's Nest," 1835

◄ Breeding song
The male yellow-breasted bunting sings a series of clear ringing notes from an elevated spot to attract a mate. Its sturdy conical bill is ideal for cracking open seeds.

Order Passeriformes

Families 1

Species 3

Size range 6–8 in (15–20 cm) long

Distribution Europe, Asia, North America

➤ **Feeding frenzy**
A hungry flock of Bohemian waxwings are drawn to the golden berries of this rowan tree, an ornamental variant that is widely planted in public spaces across the UK.

Waxwings

Bombycillidae

This distinctive family comprises just three species of beautiful and charismatic songbirds, noted for their nomadic ways and silky, ornamented plumage.

All three waxwing species breed in northerly boreal forest in loose colonies, and come winter they form large flocks (sometimes thousands strong), which roam in search of trees and shrubs that have a good berry crop. In some years, they spread thousands of miles south of their usual range. Once a flock finds a suitable food source, the birds will strip a tree of its fruit at high speed, then move on. Their digestive tract works just as efficiently: from ingesting a berry to excreting the indigestible parts takes only 16 minutes.

Roman author Pliny the Elder asserted that waxwings glow in the dark. This is not the case, but the unique red appendages that tip the secondary wing feathers of Bohemian (*Bombycilla garrulus*) and Cedar Waxwings (*B. cedrorum*) are dazzlingly bright. They are formed of highly pigmented secretions and are waxy in nature. Their size and number increase with age, and they serve as a marker of the bird's fitness and status. Like their closest relatives, the silky-flycatchers (family Ptiliogonatidae), waxwings have unusually soft plumage. Another notable trait is that, despite being songbirds, waxwings have no true song. They do, however, produce highly evocative trilling flight calls that resemble the sound of sleigh bells.

▲ **Feathers and flowers**
In this 1891 painting by Imao Keinan, a Japanese Waxwing (*B. japonica*) flies over a stand of Chinese bellflowers and other delicate wildflowers.

➤ Dominance display
An adult white-crowned sparrow aims to displace an immature bird from its perch. Many sparrows live in flocks in winter and maintain hierarchies.

PECTORAL SPARROW
Arremon taciturnus

While most members of this family are brown birds of scrub and grassland, this attractive South American species occurs in the understory of humid, tropical forest.

SPOTTED TOWHEE
Pipilo maculatus

This towhee is well known for adopting the double-scratch method of feeding on the ground, which it also uses to clear snow. In winter, its bill shortens, owing to wear from seeds and grit.

New World Sparrows

Passerellidae

Order Passeriformes

Families 1

Species More than 130

Size range 5–10 in (12–24 cm)

Distribution North, Central, and South America

Many of these seed-eating birds have brown, streaked plumage, but some can be easily distinguished by their unique song or behavior.

Unrelated to Old World sparrows, members of this family are united by having an even, conical bill for husking seeds. Most North American species are dull and brown, occur in scrub and grassland habitats, and form flocks in winter. In Central and South America, many are more boldly colored and occur in forests. A characteristic foraging method is known as double-scratching. Towhees (*Pipilo* spp.) keep their bodies still while simultaneously kicking both feet backward, as far as 6 in (15 cm), to reveal food, such as seeds and insects, hidden by leaf litter. In winter, when food is scarce, sparrows are common visitors to feeders in US gardens.

New World sparrows are territorial and sing characteristic songs. Male White-crowned Sparrows (*Zonotrichia leucophrys*) learn songs from adults and have distinct dialects. Most species are monogamous, but Saltmarsh Sparrow (*Ammospiza caudacuta*) males patrol an undefended home range and will mate with any female they meet. Members of the family generally build cup nests, but several, such as Bachman's Sparrow (*Peucaea aestivalis*), build an arch over the top for added concealment from predators.

▲ Winter visitors
Illustrated by Robert Havell c.1822, the Dark-eyed Junco (*Junco hyemalis*) or "snow-bird," is a common visitor to winter gardens.

> " . . . there is not an individual in the Union who does not know the little Snow-bird "
>
> JOHN JAMES AUDUBON, *Ornithological Biography*, 1831

New World Blackbirds

Icteridae

New World blackbirds are one of the most successful and adaptable of all songbird families. Their noisy ranks include several of the world's most abundant birds.

Order Passeriformes

Families 1

Species More than 100

Size range 6–21 in (16–53 cm) long

Distribution North America, South America

Some members of the New World blackbird family are among the best-known birds in the Americas because they are everywhere. They have adapted better than many other birds to human-created habitats and can be seen in flocks along roadsides, in city parks, and across pastureland and agricultural fields. Species such as the Red-winged Blackbird (*Agelaius phoeniceus*), Common Grackle (*Quiscalus quiscula*), Great-tailed Grackle (*Q. mexicanus*), Brown-headed Cowbird (*Molothrus ater*), and Shiny Cowbird (*M. bonariensis*) number in the many millions.

Their success is partly down to their unusual bill structure and musculature. Not only is the bill long and pointed but it can be opened with considerable force due to a modified skull. Known as "gaping," this action enables foraging birds to form holes in loose soil, open up gaps

▼ Locked in a dispute
Two male Baltimore Orioles (*I. galbula*) clash over personal space and territory.

Brown-headed cowbirds lay their eggs in the nests of more than **220 host species**

➤ **Bright-tailed bird**
Found in parts of Central America, male and female Montezuma Oropendolas (*Psarocolius montezuma*) have a very similar coloration, but males are considerably larger.

in foliage and litter, and break the skins of fruits to consume their contents. Most species are omnivorous, eating invertebrates, seeds, and fruits, but orioles (*Icterus* spp.) also forage for nectar and grackles may catch small fish.

Mating and migration

No other family of songbirds exhibits such a varied range of breeding systems. Some, such as the orioles, breed as monogamous pairs. For others, such as red-winged blackbirds, each male establishes a territory and mates with all the females that use it. Male great-tailed grackles defend a harem against other males. Yellow-rumped Marshbirds (*Pseudoleistes guirahuro*) are cooperative breeders, with adult helpers at each nest. Tricolored Blackbirds (*Agelaius tricolor*) nest in colonies of up to 100,000 nests. Some cowbird species, most notably the brown-headed cowbird, are brood parasites, laying their eggs in other species' nests to be unwittingly raised by the host parents. This often occurs at the expense of the host mother's own eggs.

Several species gather in roosts outside the breeding season and in concentrations that can exceed one million birds. And while most are nonmigratory, the Bobolink (*Dolichonyx oryzivorus*) flies from northern North America to the plains of Argentina, Bolivia, and Paraguay (with a long stop in Venezuela or Colombia) for the winter—a round trip of about 12,400 miles (20,000 km).

RED-WINGED BLACKBIRD
Agelaius phoeniceus
Male Red-winged Blackbirds with larger red patches have the advantage for attracting females. One male has been known to breed with up to 15 females in a season.

◄ **Ground nesters**
Eastern Meadowlarks (*Sturnella magna*) weave a nest from dried grasses and plant stems, usually covered with a roof. The males and females look similar, as this 1888 illustration shows.

Cardinals and Tanagers

Cardinalidae, Thraupidae

These two New World families of finchlike birds include some strikingly colorful species. The name "tanager" is derived from *tangara*, the name used by the Tupi peoples of Brazil.

The cardinals and tanagers are among the most recent of all bird groups to have evolved, appearing about 12 million years ago. Recent DNA studies have resulted in species being shuffled between passerine families and left a disproportionate number of species (about 400) in the Thraupidae family—from the lavishly colorful forest-canopy tanagers to many high Andean seed-eating species and the decidedly plain Darwin's finches. It is almost impossible to describe the core features of the family, other than the finchlike bill and relatively small size. Almost all species breed as monogamous pairs and build cup-shaped nests.

From the brilliant to the bizarre

The forest tanagers include some of the most gorgeous birds on Earth, with astonishingly bright colors, bold patterns, and iridescence. They tend to feed on fruit well above ground and, although they occur in pairs or family groups, they often form the core of mixed feeding flocks in humid tropical forests, where 10 or more species may gather. Almost all species also consume insects, and other forest tanagers, such as the vibrant honeycreepers, also feed on nectar.

Among the more modestly colored species are Darwin's finches of the Galápagos Islands. Observing and collecting these in 1835, along with the native mockingbirds (*Mimus* spp.), helped to nudge British scientist Charles Darwin toward his theory of evolution by natural selection. Their differently shaped bills correspond to their different diets, some consuming insects and others seeds of varying size. Within this group are the Woodpecker Finch (*Camarhynchus pallidus*), which breaks off twigs to use as a spear to impale grubs and pull them from a tree trunk, and the Vampire Ground-finch (*Geospiza septentrionalis*), which bites seabirds and drinks their blood.

Seed crushers

The cardinals are a much smaller family of stout-billed birds, small to medium-sized, most of which are seed-eaters, although they also may take insects and fruit. They can crush heavy seeds, and the bill of the Pyrrhuloxia (*Cardinalis sinuatus*) is curved and parrotlike. Many species are adorned with bright colors, especially blue and red. The latter includes the Northern Cardinal (*C. cardinalis*), a frequent visitor to feeders. The male has one of the finest songs in North America, while the female sings in spring, before laying her eggs.

PARADISE TANAGER
Tangara chilensis
This particularly colorful tanager is common throughout rainforests in most parts of the Amazon Basin in South America.

INDIGO BUNTING
Passerina cyanea
A popular bird in North America, the male Indigo Bunting sings its spirited song all summer long and all day long in weedy fields and scrub.

▼ Vibrant visitor
The brightly colored Golden Tanager (*Tangara arthus*) is a montane forest species, but it is also a common visitor to gardens.

➤ Beloved redbirds

Nicknamed redbird for the males' bright red plumage, the northern cardinal is one of North America's most popular birds. It has been chosen as the state bird of seven US states.

Order Passeriformes

Families 2

Species More than 440

Size range 3½–10 in (9–26 cm) long

Distribution North America, South America

A male cardinal may show aggression by flapping its wings and thrusting its head forward, toward a rival

The crest can be raised in display and is present in both males and females

The thick bill is typical for members of the cardinal family

The vibrancy of the male's red plumage varies from bird to bird and is important in attracting a mate

Northern cardinals get their **red color** from their **food**

Multicolored males
The attractive inhabitants of the state parks near Morretes, Brazil, have placed them on many a birdwatcher's travel list. Here, two male tanagers—a Blue Dacnis (*Dacnis cayana*) and a Red-necked Tanager (*Tangara cyanocephala*)—pose for the cameras above a male finch—a Violaceous Euphonia (*Euphonia violacea*)—in their native Atlantic forest habitat.

Over 35 species of warblers have been **spotted** in New York's **Central Park** during **spring migrations**

The chestnut crown patch contrasts with the bird's slate-gray head

The yellow underparts are common for the South American birds of this species; the birds in the northern part of its range have orange or red bellies

New World Warblers

Parulidae

Also known as wood warblers, members of this family of small songbirds are found throughout most of the Americas. They are not closely related to warblers elsewhere in the world.

▲ Wide-ranging whitestart
The Slate-throated Whitestart (*Myioborus miniatus*) is found from Mexico to northwestern Argentina. It has the largest range of any whitestart.

Most species of New World warblers frequent tree tops, while some inhabit the middle levels. Some species stay close to the trunks of trees, and a few birds live on the ground. Not all members of this family are restricted to woodland as some species are found in deserts or swamps.

These warblers form part of an extensive group of New World birds that all have nine visible primary feathers. Within this group, the relationships between the six families are still the subject of debate. The Parulidae family itself has evolved rapidly to produce a large number of species that are somewhat similar in appearance, giving rise to identification challenges as well as questions over the degree to which physical appearance reflects the evolutionary relationship between species. Ongoing DNA studies have overturned older classifications and still lead to frequent reshuffling within closely related species. For example, the American Yellow Warbler complex has been treated as one species or several, with as many as 43 subspecies differentiated by sometimes small differences in breeding plumage. Currently, it is considered to be one species, *Setophaga aestiva*, with nine subspecies.

Despite being called warblers, many species in this family are not noted for their singing ability and their songs are described as being more insectlike than warbling.

Colorful migrants

The New World warblers tend to have more colorful plumage than the unrelated, plainer Old World warblers. However, many tropical members of the family are dull, mostly green, and the sexes have similar plumage. In contrast, those birds that migrate to the northern temperate regions to breed famously include species with

Order Passerines

Families 1

Species 120

Size range 4–7½ in (11–19 cm) long

Distribution North and South America

◄ Presumed extinct
Bachman's warbler is depicted in this engraving by Robert Havell from Audubon's *The Birds of America* (1834). It is very likely extinct. The last confirmed sightings were all near Charleston, South Carolina, US, in 1958–61.

NORTHERN WATERTHRUSH
Parkesia noveboracensis

Breeding in northern North America, the Northern Waterthrush overwinters in Central and South America, returning to the same site year after year.

BLACKPOLL WARBLER
Setophaga striata

This species has the longest migration of any American warbler, from Alaska to Brazil. The fall migratory route includes 1,800 miles (3,000 km) over open water.

bright colorful male courtship plumages. Most notable are the species in the genus *Setophaga*, which are among the most attractive to birdwatchers. Their conspicuous spring arrival at their breeding grounds in northeastern North America is one of the most anticipated annual events for birdwatchers, and even in cities, enthusiasts search for them in gardens and parks. During the peak migration in the eastern US, more than 30 species can be seen in one day in some urban parks, of which New York City's Central Park is the most celebrated. Northwestern Ohio in the US, along the shores of Lake Erie, is another popular spot for birdwatching and the spring warbler migrations draw large crowds from all over the world. One annual event here has grown into a 10-day birdwatching festival called "The Biggest Week in American Birding."

The popularity of these warblers among birdwatchers in the US has led to the neck strain caused by staring up at birds in the canopy to be called "warbler neck." It is so common that the National Audubon Society has published an exercise routine to prepare people for the rigors of the birdwatching season and prevent repetitive strain injuries.

Foraging techniques

Most New World warblers are quite similar in shape, with a fairly short, slender bill adapted for gleaning insects from foliage. In this respect they occupy similar niches to Old World warblers. The American Redstart (*Setophaga ruticilla*) has the most flycatcherlike anatomy and behavior of any New World warbler. Indeed, it was described by

◄ Nesting supplies
The American yellow warbler nest is a deep cup built by the female, seen here collecting material from a bulrush to line the nest.

"... or a warbler a summer yellow bird makes a pretty thorough exploration about all its expanding leaflets... "

HENRY DAVID THOREAU, *Journal*, MAY 16, 1860

Linnaeus in 1766 as an Old World flycatcher named *Muscicapa ruticilla*. It is adapted to its foraging by hawking thanks to a long, flattened bill, rictal bristles, and the large surface areas of its wings and the tail.

The American redstart's tropical cousins, the whitestarts (*Myioborus* spp.), also employ the flycatching technique although not to the same extent. They spread their bright white tail feathers in order to flush insect prey, which is then deftly caught. The two waterthrushes (*Parkesia* spp.) have a very different foraging style. Their strong legs are adapted for walking on the ground, where they hunt among the leaf litter for invertebrates such as insects, spiders, and snails.

➤ Distinctive cheeks
The Red Warbler (*Cardellina rubra*) is found only in the temperate forest of three unconnected mountain ranges in Mexico. The separate populations differ in the color of the cheek patches.

The tail is duller than the body

Most New World warblers are monogamous, with the female constructing the nest but both parents raising the young. Most species build an open cup nest, located anywhere from the ground to the tree canopy.

Declining numbers

Habitat loss has affected the New World warblers' numbers. At least one fifth of the species are now of conservation concern, with 12 listed as threatened. Two species—Bachman's Warbler (*Vermivora bachmanii*) and Semper's Warbler (*Leucopeza semperi*)—are considered to be critically endangered and are likely extinct. Semper's warbler is found only on the island of St Lucia and was last seen in 1961.

The winter plumage is a pale brown for both sexes

The conspicuous yellow patch on the rump is visible whether the bird is in flight or at rest

Yaupon holly berries provide sustenance during winter

◄ Holly feast
Like many warblers, the Yellow-rumped Warbler (*Setophaga coronata*) eats insects in the breeding season, and fruit and berries in the fall and winter.

Blackpoll warblers nearly double their **body mass** before **fall migration**

Glossary

In this glossary, words in italics have a separate entry.

AIR SAC A thin-walled structure joined to the lungs of a bird, involved in respiration. Birds that plungedive into water, such as gannets and boobies, have modified air sacs beneath their skin to help cushion the impact.

ALLOPREENING Mutual *preening* between two birds, the main purpose of which is to reduce the instinctive aggression when birds come into close contact. In the breeding season, allopreening helps strengthen the pair bond between the male and female. See also *preening*

ALTRICIAL Describes young birds that hatch naked or with a patchy covering of down, often with nonfunctioning or closed eyes. They are helpless and depend entirely on their parents. See also *precocial*

ARBOREAL Living fully or mainly in trees.

AUSTRALASIA A biogeographical region that comprises Australia, New Guinea, New Zealand, and adjacent islands in the East Indies and Polynesia.

AVIAN INFLUENZA A highly infectious disease carried by birds, especially wild waterfowl and domestic chickens and turkeys. It is caused by avian influenza viruses closely related to those that cause human influenza. Also called "bird flu."

BEAK see *bill*

BILL A bird's jaws. A bill is made of bone, with a hornlike outer covering of keratin.

BINOCULAR VISION Vision in which the two eyes face forward, giving overlapping fields of view. This enables a bird to judge depth accurately and it is found mainly in predatory species.

BIRD FLU see *avian influenza*

BIRD OF PREY Any of the predatory birds in the order Accipitriformes (eagles, kites, hawks, buzzards, ospreys, and vultures), and also the orders Falconiformes (falcons and caracaras) and Strigiformes (owls). These birds are typified by their acute eyesight, powerful legs, strongly hooked *bill*, and sharp *talons*. Birds in the Accipitriformes and Falconiformes are also known as raptors.

BODY FEATHER see *contour feather*

BREEDING PLUMAGE A general term for the plumage worn by adult birds when they display and form breeding pairs. It is usually worn in the spring and summer, but this is not always the case.

BROOD PARASITE A bird that tricks another bird into raising its young. Some brood parasites always breed in this way, laying their eggs in the nests of different species. A variety of other birds are occasional brood parasites. They usually raise their own young, but sometimes dump their eggs in the nest of another.

CALL A vocal sound produced by a bird to communicate a variety of messages. Calls are often highly characteristic of individual species and can be a major aid in locating and identifying birds in the field. Most bird calls are shorter and simpler than *songs*.

CANOPY The highest layer of a forest or woodland, created by the overlapping branches of neighboring trees.

CASQUE A bony extension on the head of an animal. Cassowaries and hornbills have a casque, typically larger in the male.

CHURRING An extremely long, far-carrying, repetitive trill produced at night by some species of nightjars, as a territorial *song*.

CLOACA An opening toward the rear of a bird's body. It is present in both sexes and is used by the reproductive and excretory systems.

CLUTCH The group of eggs in a single nest, usually laid by one female and incubated by one or both parents. Clutch sizes vary from a single egg in some species, to as many as 28 eggs.

COLONY In birds, a group of the same species nesting together in the same area. Colonial nesting is common among seabirds, herons and their relatives, swifts, bee-eaters, and weavers.

COLOR FORM One of two or more clearly defined plumage variations found in the same species. Also known as a color morph or phase, a color form may be restricted to a particular part of a species' range, or it may occur side by side with other color forms throughout the entire range. Adults of different color forms are able to interbreed, and these mixed pairings can produce young of either form.

CONTOUR FEATHER A general term for any feather that covers the outer surface of a bird, including its wings and tail. Contour feathers are also known as body feathers, and help to streamline the bird.

COOPERATIVE BREEDING A breeding system in which a pair of parent birds are helped in raising their young by several other birds, which are often related to them and may be the young birds from previous broods. A few hundred species worldwide reproduce in this way.

COURTSHIP DISPLAY Ritualized, showy behaviour used in courtship by the male and sometimes also by the female.

COVERT A small feather that covers the base of a bird's flight feather. Together, the coverts form a well-defined feather tract on the wing or at the base of the tail.

CROP In birds, a muscular pouch below the throat, forming an extension to the esophagus. Its purpose is to store undigested food, and it enables birds to feed quickly so that they can digest their meal later in safer surroundings.

CRYPTIC COLORATION Coloration and markings that make an animal difficult to see against its background.

DABBLE To feed in shallow water by sieving water and food through comblike filters in the *bill*. This behavior is common in some ducks, giving rise to the name "dabbling duck."

DECURVED A term describing a bird's *bill* that curves downward toward the tip.

DIMORPHISM see *sexual dimorphism*

DIURNAL Active during the day.

DOWN FEATHER A soft, fluffy feather that provides good insulation. Young birds are covered by down feathers before they molt into their first *juvenile* plumage. Some adult birds, including waterfowl, have a layer of down feathers under their *contour feathers*.

ENDEMIC A species native to a particular geographic area, such as an island, forest, mountain, state, or country, that is found nowhere else.

FLEDGE In young birds, to leave the nest or to acquire the first complete set of flight feathers. These birds are known as fledglings, and may remain dependent on their parents for some time after fledging.

FLEDGLING see *fledge*

FLIGHT FEATHER A collective term for a bird's wing and tail feathers, used in flight. More specifically, it also refers to the largest feathers on the outer part of the wing.

GAMEBIRD A member of the order Galliformes, including species such as pheasants, quails, francolins, peafowl, turkeys, grouse, and the domestic chicken.

GAPE The mouth of a bird, or the angle at the base of its bill.

GIZZARD A muscular sac that forms the upper part of a bird's stomach. It plays an important part in grinding up food, especially in species that eat seeds and nuts, such as *gamebirds*, pigeons, doves, and parrots.

GUANO The accumulated droppings of seabirds at their nesting colony, sometimes harvested as a fertilizer.

HABITAT Any area that can support a particular group of living things.

HAWKING A feeding technique in which a bird sits motionless on a perch and waits for a flying insect to come near, then suddenly flies out to catch it in midair. Hawking is used by flycatchers and bee-eaters in particular.

IMMATURE In birds, an individual that is not yet fully mature or able to breed. Some birds, such as gulls, pass through a series of different immature plumages over a period of several years before finally adopting their first adult plumage.

INCUBATE In birds, to sit on eggs to keep them warm, allowing them to develop. Incubation is often, but not always, carried out by the adult female.

INTRODUCED SPECIES A species that humans have accidentally or deliberately brought into an area where it does not naturally occur.

IRIDESCENT PLUMAGE Plumage that shows brilliant, luminous colors, which seem to sparkle and change when seen from different angles.

JUVENILE A term referring to the plumage worn by a young bird at the time it makes its first flight and until it begins its first *molt*.

KLEPTOPARASITE A bird that gets much of its food by stealing it from other birds.

LAMELLAE Delicate, comblike structures inside the *bill* of some birds—such as flamingos, ducks, and geese—used for filtering tiny food particles out of water.

LEK A communal display area used by male birds during courtship. The same location is often revisited for many years. Species that display at leks include some *gamebirds* and *waders* and a variety of tropical forest birds, such as manakins.

MANDIBLE The upper or lower part of a bird's *bill*, known as the upper or lower mandible respectively.

MIGRATION A journey to a different region, following a well-defined route. Most birds that migrate regularly do so in step with the seasons, so that they can take advantage of good breeding conditions in one place, and good wintering ones in another.

MIMICRY In birds, the act of copying the *songs* or *calls* of other species. The mimic often weaves these vocal fragments into its own usual song. Some birds, including parrots and mynas, can also copy mechanical sounds such as ringing telephones, machinery, and car alarms, and even human speech.

MOBBING A type of defensive behavior in which a group of birds gang up to harass a predator, such as a *bird of prey*, swooping repeatedly to drive it away.

MONOGAMOUS Mating with a single partner, either in the course of a single breeding season or for life. See also *polygamous*

MORPH see *color form*

MOLT In birds, to shed old feathers so they can be replaced. Molting enables birds to keep their plumage in good condition, to change their level of insulation, and to change their coloration or markings when they are ready to breed.

NEW WORLD The Americas, including the Caribbean and offshore islands in the Pacific and Atlantic oceans.

OLD WORLD Europe, Asia, Africa, and Australasia. See also *New World*.

PELAGIC Relating to the open ocean. Pelagic birds spend most of their life at sea and only come to land to nest.

PHASE see *color form*

POLYGAMOUS Mating with two or more partners during the course of a single breeding season. See also *monogamous*

PRECOCIAL Describing young birds that are well developed at hatching, with down feathers and open, functioning eyes. Many precocial young are soon able to walk or swim and find food themselves. See also *altricial*

PREENING Essential routine behavior by which birds keep their feathers in good condition. A bird grasps an individual feather at its base and then "nibbles" upwards toward the tip, and repeats the process over and over with different feathers. This helps to smooth and clean the plumage. Birds often also smear oil from their preen gland onto their feathers at the same time. See also *allopreening*

PRIMARY FEATHER One of the large outer wing feathers, growing from the digits of a bird's "hand." The primary feathers are often collectively referred to as primaries. See also *secondary feather*

ROOST A place where birds sleep, or the act of sleeping. The majority of birds roost at night. However, nocturnal species such as owls and nightjars roost by day, and many coastal species, particularly *waders*, roost at high tide when their feeding areas are flooded by the rising seawater.

SALLY A short flight from a perch to catch an invertebrate, often in midair. See also *hawking*

SCRAPE A simple nest that consists of a shallow depression in the ground, which may be unlined or lined with soft material such as feathers and grasses.

SECONDARY FEATHER One of the row of long, stiff feathers along the rear edge of a bird's wing, between the body and the primary feathers at the wingtip. The secondary feathers are often collectively referred to as secondaries. See also *primary feather*

SEXUAL DIMORPHISM The occurrence of physical differences between males and females of a species. In birds, the most common type of sexual dimorphism is plumage variation. Other forms of sexual dimorphism include differences in bill length or body size.

SHOREBIRD see *wader*

SONG A loud vocal performance by a bird, usually the adult male, to attract and impress a potential mate, advertise ownership of a *territory*, or drive away rival birds. Songs are often highly characteristic of individual species and can be a major aid in locating and identifying birds in the field. See also *call*

SYRINX A modified section of a bird's trachea (windpipe), equivalent to the voice box in humans, that enables birds to call and sing.

TALON The sharp, hooked claw of a *bird of prey*.

TERRITORY An area defended by an animal, or group of animals, against other members of the same species. Territories often include useful resources, such as good breeding sites or feeding areas, which help a male attract a mate. In birds, territories vary in size from just a few inches wide in colonial species to many square miles in some large eagles.

TUBENOSE A general term used informally to describe members of the order Procellariiformes, including albatrosses, petrels, and shearwaters. It refers to the distinctive tubular nostrils on their upper bill, which are not found in any other birds.

VESTIGIAL An organ or biological structure that is retained, despite seeming to have lost its original function.

WADER Any member of several families in the order Charadriiformes. Waders typically have a long bill and long legs. Many of them occur at the water's edge in wetlands or along coasts, but despite their name not all species actually wade in water and some live in quite dry habitats. An alternative name for waders is shorebirds, especially in the US and Canada.

WATTLE A bare, fleshy growth that hangs loosely below the bill in some birds. It is often brightly colored, and may play a part in courtship.

WATERFOWL A collective term for members of the order Anseriformes, including screamers, ducks, geese, and swans.

WILDFOWL A collective term for ducks, geese, and swans (family Anatidae).

WINGSPAN The distance across a bird's outstretched wings and back, from one wingtip to the other.

ZYGODACTYL FEET A specialized arrangement of the feet in which the toes are arranged in pairs, with the second and third toes facing forward and the first and fourth toes facing backward. This adaptation helps birds to climb and to cling to tree trunks and other vertical surfaces.

Index

Page numbers in **bold** refer to main entries.

A

B

D

E

M

N

O

P

QR

S

T

UVW

Acknowledgments

DK would like to thank the following people for their help with making this book:

Proofreading: Salima Hirani

Indexing: Helen Peters

Fact-checking: Angela Modany, Michelle Rae Harris

Illustrator: Phil Gamble

Design assistance: Francis Wong

Technical assistance: Sonia Charbonnier, Tommy Callan

Picture research: Kate Sayer, Amy English, Martin Copeland, Romaine Werblow

Color retouching: Steve Crozier

Help with cutouts: Pankaj Sharma, Jagtar Singh

For **Smithsonian Enterprises:**

Licensing Coordinator: Avery Naughton,

Editorial Lead: Paige Towler

Senior Director, Licensed Publishing: Jill Corcoran

Vice President of New Business and Licensing: Brigid Ferraro

President: Carol LeBlanc

The publisher would like to thank the following for their kind permission to reproduce their photographs:

(Key: a-above; b-below/bottom; c-center; f-far; l-left; r-right; t-top)

1 Dorling Kindersley: Richard Leeney / Maidstone Museum and Bentliff Art Gallery. **2-3 Dorling Kindersley:** Peter Chadwick / Natural History Museum, London (c). **Dreamstime.com:** Evgeniya Moroz (bc); Picstudio (ca/kingfisher feather). **Getty Images / iStock:** DrPAS (cb); E+ / Gregory_DUBUS (ca). **2 123RF.com:** mycteria (cra/gray featherx2). **Dorling Kindersley:** Peter Chadwick / Natural History Museum, London (tr). **Getty Images / iStock:** mycteria (cr); nadtytok (crb). **3 Dorling Kindersley:** (crb/hawk feather); Philip Dowell / Natural History Museum, London (bc/red feather); Barnabas Kindersley (tr); Peter Chadwick / Natural History Museum, London (bl, bc/yellow feathers, cra/Cockatiel, cra/Gull feather, cra/macaw feather, tc, cla, tl/macaw feather, cla/Owl feather). **Dreamstime.com:** Olga Bobrovskaya (bc, cra, tc/yellow feathers); Wollertz (crb); Natallia Yaumenenka (tr/green feather); Smileus (ca). **Getty Images / iStock:** DrPAS (crb/orange feather); mycteria (br, cr). **Shutterstock.com:** Gallinago_media (tl). **4-5 Alamy Stock Photo:** Cynthia Lee (t). **6-7 Dreamstime.com:** Irina Iarovaia. **7 Deepak Kumar:** (br). **8-9 Photo Scala, Florence:** The Metropolitan Museum of Art / Art Resource. **10-11 naturepl.com:** Richard Flack. **Shutterstock.com:** Dn Br (background). **13 Getty Images / iStock:** igorkov (tr); KenCanning (tl). **14 Alamy Stock Photo:** Pictorial Press Ltd (bl); Ivan Vdovin (cl). **© The Trustees of the British Museum. All rights reserved:** (cla). **Dorling Kindersley:** Andy Crawford / Senckenberg Nature Museum (br). **15 Alamy Stock Photo:** All Canada Photos / Wayne Lynch (br); GRANGER – Historical Picture Archive (cb); Classic Image (clb). **Artwork based on a photo by Professor Daniel J. Field, University of Cambridge:** (cra) **Dorling Kindersley:** Jon Hughes (crb). **Ksepka, D., L. Grande, and G. Mayr. 2019. Oldest finch-beaked birds reveal parallel ecological radiations in the earliest evolution of passerines. Current Biology 29 (4): 657-663:** (cla). **16 123RF.com:** Lynn Bystrom / critterbiz (br). **Dreamstime.com:** Sue Feldberg / Suefeldberg (c); Randall Runtsch (c/robin). **Getty Images / iStock:** PrinPrince (cr). **17 Dreamstime.com:** K Quinn Ferris (c/blue jay); Ondej Prosick (c); Jan Martin Will (br). **19 Alamy Stock Photo:** All Canada Photos / Glenn Bartley (t). **Dorling Kindersley:** Frank Greenaway (b). **21 Alamy Stock Photo:** James Crooke (t). **Getty Images / iStock:** Serhii Luzhevskyi (br). **22 Getty Images / iStock:** E+ / Andyworks (cr); MriyaWildlife (c). **Getty Images:** Stockbyte / Andrew Howe (cl). **Harprit Singh:** (br). **23 Getty Images / iStock:** Marc Andreu (cl); Mark Kostich (c); ozflash (cr); Sander Meertins (br). **24 Alamy Stock Photo:** Everett Collection Historical (cl); IanDagnall Computing (cb); JJs (bl); Shawshots (bc). **Dreamstime.com:** Daniel Prudek (cra). **Getty Images:** David Lefranc / Gamma-Rapho (crb). **25 Alamy Stock Photo:** Penta Springs Limited (crb); World History Archive / Desmond Morris Collection (cla); The Picture Art Collection (cr). **Dreamstime.com:** Stig AlenA?s (cra). **Getty Images:** Hulton Archive / Heritage Images (cl). **naturepl.com:** David Tipling (tl). **26 Alamy Stock Photo:** Mint Images Limited (bl). **Dreamstime.com:** Wkruck (tr). **27 Alamy Stock Photo:** William Graham (cra); Historic Illustrations / PhotoStock-Israel (tc); Tammy Wolfe (br). **Dreamstime.com:** Bernard Breton (bc); Engl01 (bl). **28 Biodiversity Heritage Library:** flickr / n516_w1150 (br). **29 Getty Images / iStock:** GlobalP (tr). **30-31 Getty Images:** De Agostini Picture Library. **Getty Images / iStock:** E+ / tracielouise (Cockatoo). **32 Alamy Stock Photo:** Johan Swanepoel. **32-33 Getty Images / iStock:** Hein Nouwens (background). **33 Bridgeman Images:** © NPL – DeA Picture Library / S. Vannini (tr). **© The Trustees of the British Museum. All rights reserved:** (crb). **34 123RF.com:** Eric Isselee (t). **Getty Images / iStock:** DigitalVision Vectors / ilbusca. **35 Getty Images / iStock:** Cormorant (cr); Hein Nouwens; Ken Griffiths (bl). **36-37 Shutterstock.com:** Intellegent Design (background). **Unsplash:** Zdenk Machek. **37 Bridgeman Images:** Alinari (tr). **Dorling Kindersley:** Gary Ombler / University of Aberdeen (br). **38 Dreamstime.com:** Wirestock (bl). **38-39 Alamy Stock Photo:** Corine van Kapel (t). **Shutterstock.com:** Intellegent Design (backgroung). **39 Dreamstime.com:** Isselee (crb); Alexander Potapov (cra). **40-41 Shutterstock.com:** Intelligent Design (background). **40 Getty Images / iStock:** mirceax (br). **41 Alamy Stock Photo:** Corine van Kapel (t). **Bridgeman Images:** Granger (br). **42-43 Terje Kolaas. 44 Alamy Stock Photo:** Art Collection 3 (tl). **44-45 Alamy Stock Photo:** Quagga Media. **Getty Images:** 500px Asia / Zhu Dianhua (t). **46 Alamy Stock Photo:** VTR (l). **46-47 Alamy Stock Photo:** Quagga Media. **47 Dreamstime.com:** Jeffrey Banke (br); Isselee (cr). **naturepl.com:** Gerrit Vyn (tl). **48 Alamy Stock Photo:** Fabrice Bettex Photography (cr); Florilegius (bl). **48-49 Getty Images / iStock:** OLHA POTYSIEVA (background). **49 Dorling Kindersley:** E. J. Peiker (cr); Andy and Gill Swash (crb). **Getty Images:** imageBROKER / Marko Konig (cb). **50 The Metropolitan Museum of Art:** Gift of Helen Miller Gould, 1910 (tl). **50-51 Alamy Stock Photo:** Nature Picture Library / Claudio Contreras (b). **Dreamstime.com:** Evgeny Turaev. **52-53 Alamy Stock Photo:** Wirestock, Inc.. **54 Alamy Stock Photo:** AGAMI Photo Agency / Daniele Occhiato (t); The Natural History Museum (bl). **Dorling Kindersley:** George Lin (crb). **55 Alamy Stock Photo:** DBI Studio (bl). **Dreamstime.com:** Markusmayer1. **Shutterstock.com:** Holger Kirk (br). **56 Getty Images:** Moment Open / Picture by Tambako the Jaguar. **56-57 Alamy Stock Photo:** GRANGER – Historical Picture Archive. **57 Brooklyn Museum:** Gift of Mary Babbott Ladd, Lydia Babbott Stokes, and Frank L. Babbott, Jr. in memory of their father Frank L. Babbott (tr). **58 Alamy Stock Photo:** GRANGER – Historical Picture Archive; North Wind Picture Archives (bl). **Dreamstime.com:** Dennis Jacobsen (cra). **Getty Images / iStock:** Gerald Corsi (crb/Dove); E+ / Whiteway (crb). **Brad Imhoff:** (tc). **59 The New York Public Library:** Rare Book Division, The New York Public Library. "Syrrhaptes paradoxus. Pallas's Sand-Grouse." The New York Public Library Digital Collections. 1862 - 1873. https://digitalcollections.nypl.org/items/510d47d9-7101-a3d9-e040-e00a18064a99 (b). **Jugal Tiwari:** (cr). **60 Shutterstock.com:** Rudmer Zwerver. **61 Bridgeman Images:** From the British Library archive (tr). **Dorling Kindersley:** Ian Montgomery / Birdway.com.au (bc). **Dreamstime.com:** Paul Mckinnon (bl). **Getty Images / iStock:** Marc Guyt (br). **62 Dorling Kindersley:** Cotswold Wildlife Park & Gardens, Oxfordshire, UK (cl). **Dreamstime.com:** Steve Allen (clb). **Getty Images / iStock:** tane-mahuta (r). **62-63 Shutterstock.com:** AlinArt (background). **63 Alamy Stock Photo:** Florilegius (bc). **The Art Institute of Chicago:** Purchased with funds provided by Mrs. James W. Alsdorf (t). **64 Alamy Stock Photo:** Franka Slothouber / NIS / Minden Pictures (t). **naturepl.com:** Jack Dykinga (crb). **64-65 Dreamstime.com:** Markusmayer1 (background). **65 Alamy Stock Photo:** The Picture Art Collection (l). **Dreamstime.com:** Carol Hancock (crb); Prin Pattawaro (cra). **66-67 Alamy Stock Photo:** Authentic-Originals. **66 Shutterstock.com:** imagevixen (b). **67 Alamy Stock Photo:** The Picture Art Collection (tr). **Getty Images:** Sylvain Cordier / Gamma-Rapho (br). **68 Alamy Stock Photo:** Agami / Helge Sorensen (br); blickwinkel / M. Woike (ca). **Dreamstime.com:** Agami Photo Agency (tr). **68-69 Getty Images / iStock:** Brandy Kubig. **69 Bridgeman Images:** © Look and Learn (br). **Alfred Yan:** (t). **70 Bridgeman Images:** Prismatic Pictures. **70-71 Alamy Stock Photo:** arxichtu4ki (background). **71 Alamy Stock Photo:** William Leaman (br). **Getty Images:** Image by David G Hemmings (ca). **72 Alamy Stock Photo:** blickwinkel / M. Woike (tl). **Lynn Griffiths:** (cl). **William J Pohley:** PohleyPhoto.com (bl). **72-73 Alamy Stock Photo:** arxichtu4ki (background); Glenn Bartley / All Canada Photos (ca). **73 Alamy Stock Photo:** Erik Schlogl (crb). **74-75 Getty Images / iStock:** Christine Kohler (background). **74 Rijksmuseum, Amsterdam. 75 Shutterstock.com:** Sourav Makhal (t). **76-77 Getty Images / iStock:** Christine Kohler (background). **76 Alamy Stock Photo:** blickwinkel (tl). **Dreamstime.com:** Feathercollector (br); Glenn Price (bc). **77 Collection of Christchurch Art Gallery Te Puna o Waiwhet. purchased 2014:** John Gould and Henry Richter / Notornis Mantelli [Takah] (b). **Getty Images / iStock:** MikeLane45 (tr). **78-79 Shutterstock.com:** Vladimir Kogan Michael. **University of Nebraska-Lincoln:** Copyright Paul A. Johnsgard; courtesy Zea Books (background). **79 Alamy Stock Photo:** The Picture Art Collection (br). **80 Getty Images:** Future Publishing (t). **Reproduced by permission of RSPB, © 2024 All rights reserved:** (br). **80-81 Alamy Stock Photo:** Markus Mayer (background). **81 Dreamstime.com:** Myscamper (cr); Martin Pelanek (tr). **Getty Images:** Dejan Zakic / 500px (c). **82 Alamy Stock Photo:** Natural History Museum, London (t). **82-83 Getty Images / iStock:**

DigitalVision Vectors / NNehring. **83 Alamy Stock Photo:** Nature Picture Library / Gerrit Vyn (tl); Nature Picture Library / Mark Hamblin / 2020VISION (br). **Dorling Kindersley:** David Cottridge (cra). **84-85 Alamy Stock Photo:** Arterra Picture Library / Arndt Sven-Erik. **Getty Images / iStock:** DigitalVision Vectors / GeorgePeters (background). **85 Alamy Stock Photo:** Nature Picture Library / Roger Powell (tr). **86 Alamy Stock Photo:** David Tipling Photo Library (tl); Nature Photographers Ltd / Paul R. Sterry (crb). **Dorling Kindersley:** Mike Lane (crb/Pectoral). **Getty Images / iStock:** DigitalVision Vectors / GeorgePeters. **Rijksmuseum, Amsterdam:** Gift of J. Perre, Eindhoven (bl). **87 Biodiversity Heritage Library:** Flickr / n396_w1150 (br). **Getty Images / iStock:** Andy_Mayes_Africa (tr). **88 Dorling Kindersley:** Windrush Photos / David Tipling (clb). **Dreamstime.com:** Wirestock (bl). **88-89 Alamy Stock Photo:** All Canada Photos / Wayne Lynch. **Dreamstime.com:** Patrick Guenette (background). **89 Alamy Stock Photo:** Library Book Collection (tr). **Shutterstock.com:** Simon C. Stobart (crb). **90 Shutterstock.com:** tryton2011 (tl). **90-91 Dreamstime.com:** Evgeny Turaev (background). **91 Getty Images:** Scott Suriano. **92 Dorling Kindersley:** E. J. Peiker (cla). **Dreamstime.com:** Tarpan (cl, bl). **Getty Images:** Kevin Schafer (r). **92-93 Dreamstime.com:** Evgeny Turaev (background). **93 Alamy Stock Photo:** Hugh Harrop / AGAMI Photo Agency (c). **Biodiversity Heritage Library:** (tr). **94 Getty Images:** Moment / Pone Pluck. **94-97 Getty Images / iStock:** Stefan_Alfonso (Background). **95 Shutterstock.com:** Uzzal Kumar Kundu (ca). **96 Alamy Stock Photo:** Old Images (tl). **Dreamstime.com:** Rachel Hopper (clb); Isselee (cl). **Shutterstock.com:** Hansie Oosthuizen (bl). **97 © The Trustees of the British Museum. All rights reserved.:** (ca). **Getty Images:** Nick Dale / Design Pics (b). **98 Dreamstime.com:** Jiri Fejkl (b). **99 Alamy Stock Photo:** Addictive Stock (crb); Album (tr); All Canada Photos / Glenn Bartley (br). **Dreamstime.com:** Tarpan (cra). **Science Photo Library:** Nature Picture Library / John Shaw (bl). **100 Shutterstock.com:** Roger ARPS BPE1 CPAGB. **100-103 Getty Images / iStock:** DigitalVision Vectors / Stefan_Alfonso (Background). **101 Alex Correia:** (t). **Science Photo Library:** Natural History Museum, London (crb). **102 naturepl.com:** Doug Gimesy (b). **103 Alamy Stock Photo:** SuperStock / Ralph Lee Hopkins (tl). **Dorling Kindersley:** Cotswold Wildlife Park / Gary Ombler (cr); Richard Leeney / Whipsnade Zoo (crb). **104-105 Tim Flach. 106-107 Alamy Stock Photo:** Markus Mayer. **106 Getty Images / iStock:** E+ / Byronsdad (c). **107 Alamy Stock Photo:** Michel & Gabrielle Therin-Weise (cb). **Bridgeman Images:** Granger (tr). **Getty Images / iStock:** mantaphoto (crb/Waved Albatross); mauinowl (crb). **108 Biodiversity Heritage Library:** Flickr / Ornithologicalm2_0150 (cr). **Getty Images / iStock:** Wirestock (bl). **108-109 Getty Images / iStock:** MorePics (b). **109 Alamy Stock Photo:** Minden Pictures / Jan Vermeer (l). **Dreamstime.com:** Agami Photo Agency (cr, crb); Frank Gnther (cra). **110-111 Dreamstime.com:** Stu Porter (t). **Getty Images / iStock:** MorePics (background). **111 Alamy Stock Photo:** Chronicle (crb). **Getty Images / iStock:** DigitalVision Vectors / Grafissimo (tr). **112 123RF.com:** Ondrej Prosicky (br). **Dreamstime.com:** Mohamad Khairi Raimi (bc); Wrangel (bl). **Getty Images:** imageBROKER / Justus de Cuveland (tl). **112-113 Getty Images / iStock:** MorePics. **113 Getty Images / iStock:** SoumenNath (r). **114 Alamy Stock Photo:** Nature Picture Library / David Tipling (tr). **Michael Poliza Photography:** (br). **114-115 Dreamstime.com:** Markusmayer1. **115 Getty Images:** Stone / Ignacio Palacios (br). **116 Alamy Stock Photo:** Nature Picture Library / SCOTLAND: The Big Picture (t). **116-117 Shutterstock.com:** Morphart Creation. **117 Alamy Stock Photo:** Chronicle (tr). **Dorling Kindersley:** Jens Eriksen,Hanne (crb). **Dreamstime.com:** Zhbampton (cra). **Shutterstock.com:** Michael Nolan / Splashdown (bl). **118-119 Getty Images:** Silver Stone / 500px. **119 123RF.com:** wmarissen (crb). **Alamy Stock Photo:** BTEU / RKMLGE (tr). **Dreamstime.com:** Chris Kruger (cra). **120-121 Alamy Stock Photo:** Inge Johnsson. **122-123 Getty Images / iStock:** John_Wijsman. **122 Tracey Lund. 123 Alamy Stock Photo:** imageBROKER.com GmbH & Co. KG / David & Micha Sheldon (cra). **Bridgeman Images:** Prismatic Pictures (b). **Getty Images / iStock:** 3dotsad (crb). **124 The Metropolitan Museum of Art:** The Jefferson R. Burdick Collection, Gift of Jefferson R. Burdick (cr). **naturepl.com:** Cyril Ruoso (tr). **125 Alamy Stock Photo:** imageBROKER.com GmbH & Co. KG / Nigel Dennis (tr); robertharding / James Hager (bl). **Getty Images / iStock:** DigitalVision Vectors / GeorgePeters. **126-129 Shutterstock.com:** Artinblackink. **126 Alamy Stock Photo:** WorldFoto (t). **127 Dreamstime.com:** Ildiko Laskay. **128 Alamy Stock Photo:** All Canada Photos / Glenn Bartley (b). **The Cleveland Museum Of Art:** (tr). **129 Callie de Wet:** (bc). **Dorling Kindersley:** George McCarthy (cr). **Dreamstime.com:** Steve Byland (br); Martin Maritz (tr). **Library of Congress, Washington, D.C.:** LC-DIG-ppmsca-27739 (t). **130-131 Rawpixel. 132-133 Getty Images / iStock:** rraya (b). **Shutterstock.com:** HuanPhoto (t). **133 Alamy Stock Photo:** Mathias Putze (br). **134 Dreamstime.com:** Agami Photo Agency (bc); Riaanvdb (bl); FlorianAndronache (br); Johncarnemolla (cla). **134-135 Getty Images / iStock:** JackVandenHeuvel; rraya (background). **136 Alamy Stock Photo:** Chronicle (bl); Image Professionals GmbH / Konrad Wothe (br). **Getty Images / iStock:** Val_Iva. **137 Dreamstime.com:** Anna C. Nagel (bl). **Getty Images / iStock:** barbaraaaa (crb); Andrii-Oliinyk. **138-139 Dreamstime.com:** Volodymyr Byrdyak. **138 Getty Images:** Luis Robayo / AFP (t). **139 Alamy Stock Photo:** ZSSD / Minden Pictures (tr). **Dreamstime.com:** Edurivero (crb/Black Vulture); Victor Zherebtsov (cra); Peter Erik Millenaar (crb); Miroslav Liska (bl). **140 Shutterstock.com:** Wang LiQiang. **140-141 Getty Images / iStock:** DigitalVision Vectors / sam_ding. **141 Dorling Kindersley:** Gary Ombler / University of Pennsylvania Museum of Archaeology and Anthropology (bc). **Getty Images:** DEA / G. DAGLI ORTI / De Agostini (tr). **142-143 Getty Images / iStock:** DigitalVision Vectors / sam_ding. **142 Science Photo Library:** Pascal Goetgheluck (bc). **Shutterstock.com:** Hilton Chen / Solent News (t). **143 Alamy Stock Photo:** Penta Springs Limited / Artokoloro (tr). **Getty Images / iStock:** Andyworks (b). **144-145 Getty Images / iStock:** DigitalVision Vectors / sam_ding. **Shutterstock.com:** Marvelous Art (t). **144 Alamy Stock Photo:** Edwin Remsberg (br). **Dreamstime.com:** Dennis Jacobsen (bc). **Shutterstock.com:** Imogen Warren (bl). **145 Alamy Stock Photo:** Antiquarian Images (tr). **Getty Images:** Avalon / Universal Images Group (b). **146-147 Terje Kolaas. 148 akg-images:** Francis Dzikowski. **148-151 ArtydoveDesigns (via Etsy). 149 Getty Images / iStock:** GlobalP (br). **150-151 Getty Images:** Daniel Parent (t). **150 Alamy Stock Photo:** Douglas Lander (br). **151 Alamy Stock Photo:** maheshbhai vala (br). **Dorling Kindersley:** Frank Greenaway / Natural History Museum, London (tl). **Dreamstime.com:** Stephen Mcsweeny (tc). **152 Getty Images:** Florilegius / Universal Images Group (bl); Moment / © Juan Carlos Vindas (r). **Getty Images / iStock:** Val_Iva. **153 Alamy Stock Photo:** ART Collection (crb); imageBROKER.com GmbH & Co. KG / Thomas Hinsche (tl). **154-155 Getty Images:** Philippe Clement / Arterra / Universal Images Group (c). **154 Alamy Stock Photo:** Doug McCutcheon (bc). **Horniman Museum and Gardens:** (tr). **155 123RF.com:** carinablofield (crb); feathercollector (cr). **156 Dreamstime.com:** Kajornyot (b). **156-157 Getty Images / iStock:** Grafissimo / DigitalVision Vectors (background). **157 Dorling Kindersley:** Richard Leeney / Maidstone Museum and Bentliff Art Gallery (br). **Getty Images:** Michael Kalika / 500px (ca). **158 Dreamstime.com:** Henkbogaard (bl); Martin Pelanek (bc). **naturepl.com:** Donald M. Jones (cla). **Shutterstock.com:** Ken Griffiths (br). **158-159 Getty Images / iStock:** Grafissimo / DigitalVision Vectors (background). **159 Alamy Stock Photo:** Heritage Image Partnership Ltd (cr). **160-161 Getty Images / iStock:** mauribo. **162-163 Alamy Stock Photo:** Les Archives Digitales (background). **Getty Images:** 500Px Plus / Marco Redaelli. **162 Dreamstime.com:** Andrey Gudkov (tl). **164 Alamy Stock Photo:** Kit Day (tl); Nature Picture Library / Anup Shah (bc). **164-165 Alamy Stock Photo:** Les Archives Digitales. **165 Alamy Stock Photo:** imageBROKER.com GmbH & Co. KG / Christian Htter (cr); Nate Chappell / BIA / Minden Pictures (tr). **Bridgeman Images:** (bc). **Dreamstime.com:** Gopause (crb). **166 Alamy Stock Photo:** Alf Jacob Nilsen (tr). **Getty Images / iStock:** billberryphotography (clb). **167 Alamy Stock Photo:** Florilegius (br). **Edward Selfe:** (t). **Shutterstock.com:** Babich Alexander. **168-169 Getty Images / iStock:** duncan1890. **168 Alamy Stock Photo:** Debapratim Saha / BIA / Minden Pictures (cb). **169 Biodiversity Heritage Library:** Flickr / n434_w1150 (tl). **Dreamstime.com:** Henkbogaard (tr); Kaido Rummel (cr); Yadvendra Kumar (br). **Shutterstock.com:** Agami Photo Agency (crb). **170-171 Getty Images / iStock:** Val_Iva (background). **170 Rick Dunlap. 171 Alamy Stock Photo:** All Canada Photos / Glenn Bartley (crb); Vitalli (cb). **SuperStock:** Andrew McLachlan / All Canada Photos (cra); Photoshot (tr). **172 Dreamstime.com:** Shubhrojyoti Datta (tl). **Getty Images:** Werner Forman / Universal Images Group (br). **173 Alamy Stock Photo:** bilwissedition Ltd. & Co. KG. **174 Dreamstime.com:** Suerob (bc). **Getty Images / iStock:** Irving A Gaffney (tr). **Shutterstock.com:** Don Mammoser (bl); Sacharewicz Patryk (br). **175 Getty Images:** Moment Open / Alice Cahill (r). **176 Biodiversity Heritage Library:** Flickr / n118_w1150 (tr). **naturepl.com:** Richard Du Toit (tl). **177 Dreamstime.com:** Markusmayer1. **Getty Images / iStock:** GlobalP (br). **naturepl.com:** Pete Oxford (bl). **178-179 Getty Images / iStock:** DigitalVision Vectors / Nastasic. **178 Getty Images:** Boris Droutman / 500px (c). **179 SuperStock:** Iberfoto Archivo (t). **180-181 Getty Images / iStock:** DigitalVision Vectors / Nastasic. **180 Bridgeman Images:** (tl). **Dreamstime.com:** Assoonas (bc); Donyanedomam (bl). **Shutterstock.com:** Jonathan Chancasana (br). **181 Alamy Stock Photo:** Blue Planet Archive SKO (bl). **182-183 Getty Images / iStock:** Parrotstarr (c); Val_Iva. **182 Bridgeman Images:** (bl). **184 Dorling Kindersley:** Coppola Studios. **185 Alamy Stock Photo:** Robert Wyatt (bc). **Getty Images / iStock:** Bkamprath (crb); PrinPrince (cr). **Shutterstock.com:** Agami Photo Agency (tc). **186-187 Dr. K. M. Anand. 188-189 Getty Images / iStock:** DigitalVision Vectors / ilbusca. **188 Dreamstime.com:** Thawats (bl, bc). **naturepl.com:** Steven David Miller (br). **Shutterstock.com:** Panu Ruangjan (clb). **189 Getty Images:** De Agostini / DEA / A. DE GREGORIO. **190-191 Getty Images / iStock:** Yuliia Khvyshchuk. **190 Alamy Stock Photo:** Tom Friedel / AGAMI Photo Agency (bc). **191 Alamy Stock Photo:** Glenn Bartley / All Canada Photos (tr); Luiz Claudio Marigo / Nature Picture Library (bc). **Dreamstime.com:** Gabriel Rojo (bl). **Shutterstock.com:** Rob Jansen (br). **192 Alamy Stock Photo:** robertharding / G&M Therin-Weise (t). **Biodiversity Heritage Library:** Flickr / n598_w1150 (crb). **Shutterstock.com:** Morphart Creation. **193 Alamy Stock Photo:** David Tipling Photo Library (br); Octavio Campos Salles (tr). **194 Alamy Stock Photo:** The History Collection (bl). **194-195 Shutterstock.com:** AlinArt. **195 Alamy Stock Photo:** AGAMI Photo Agency / Dubi Shapiro (cra); Kit Day (cr); Gabbro (bc); BIOSPHOTO / scar Dez Martnez (br). **Dreamstime.com:**

Agami Photo Agency (tc). **196 Alamy Stock Photo:** Alan Murphy / BIA / Minden Pictures. **197 Alamy Stock Photo:** Authentic-Originals. **Biodiversity Heritage Library:** Flickr / n211_w1150 (tr). **198-199 Alamy Stock Photo:** Authentic-Originals. **Getty Images:** Moment / Carlos Carreno (b). **SuperStock:** Claudio Gonzales Rodriguez / BIA / Minden Pictures (tc). **199 Dreamstime.com:** Bouke Atema (cra); Ricardo De Paula Ferreira (cr). **Getty Images / iStock:** Chase D'animulls (br). **200 Biodiversity Heritage Library:** Flickr / n63_w1150_300_RT2. **200-201 Depositphotos Inc:** Morphart. **201 Jono Dashper:** (tr). **NPWS logo used with permission of NSW National Parks and Wildlife Service:** (b). **202-203 Shutterstock.com:** Bodor Tivadar. **202 naturepl.com:** Etienne Littlefair (b). **203 Alamy Stock Photo:** Natural History Museum, London (tr). **Shutterstock.com:** Wright Out There (crb). **SuperStock:** Animals Animals (cr). **204 Biodiversity Heritage Library:** Flickr / n99_w1150 (tl). **204-205 Shutterstock.com:** AlinArt. **205 Alamy Stock Photo:** Gabbro (crb); Georgie Greene (tr). **Getty Images / iStock:** Andrew Haysom (b). **naturepl.com:** Martin Willis (cr). **206 Alamy Stock Photo:** Agami / Georgina Steytler (ca). **naturepl.com:** Neil Fitzgerald / BIA (br). **206-207 Alamy Stock Photo:** Authentic-Originals. **207 Alamy Stock Photo:** blickwinkel / AGAMI / D. Shapiro (cr). **Biodiversity Heritage Library:** Flickr / n62_w1150 (ca). **Dreamstime.com:** Isonphoto (br). **208-209 Shutterstock.com:** DuyMy (t). **208-211 Getty Images / iStock:** imenachi. **209 naturepl.com:** Martin Willis (b). **210 Alamy Stock Photo:** AGAMI Photo Agency / Dubi Shapiro. **211 Biodiversity Heritage Library:** Flickr / n19_w1150 (tl). **Dreamstime.com:** Imogen Warren (br). **Getty Images / iStock:** Eric Middelkoop (tr). **212 Bridgeman Images:** From the British Library archive (tr). **Michael Riffel:** via Macaulay Library (bl). **212-213 Alamy Stock Photo:** Nature Picture Library / Dave Watts. **213 Dreamstime.com:** Steve Byland (br); Panuruangjan (crb). **214-215 Alamy Stock Photo:** Authentic-Originals. **214 Deepak Kumar:** (b). **215 Alamy Stock Photo:** Agami / James Eaton (tr); SB Stock (cra); RooM the Agency / kristianbell (br). **216 Alamy Stock Photo:** Media Drum World / Liu Chia-Pin (tl). **Getty Images / iStock:** ilbusca. **naturepl.com:** Staffan Widstrand / Wild Wonders of China (r). **217 Alamy Stock Photo:** Rick & Nora Bowers (t). **Dreamstime.com:** Imogen Warren (clb). **Getty Images / iStock:** Darya Alekseyuk. **Shutterstock.com:** Wright Out There (cl). **218-219 Getty Images / iStock:** Hein Nouwens. **218 Alamy Stock Photo:** Heritage Image Partnership Ltd (l); Zoonar / Siegmar Tylla (br). **219 Alamy Stock Photo:** Nature Picture Library / Roger Powell (tl). **Shutterstock.com:** feathercollector (cr); ZakiFF (crb). **220 Alamy Stock Photo:** Science Photo Library / PHOTOSTOCK-ISRAEL (tc). **220-223 Getty Images / iStock:** invincible_bulldog (background). **220-221 Alamy Stock Photo:** Media Drum World / Dustin Chen. **222 Dreamstime.com:** Chansom Pantip (bl). **naturepl.com:** Tim Laman (clb). **223 Alamy Stock Photo:** Nature Picture Library / David Tipling (tr); Nature Picture Library / Tim Laman (br). **naturepl.com:** Tim Laman (bc); Shane P. White (bl). **224 Shutterstock.com:** Swarnendu Chatterjee / Solent News (t). **224-225 Alamy Stock Photo:** Authentic-Originals. **225 Dreamstime.com:** Agami Photo Agency (cr, br); Martingraf (tc); Prin Pattawaro (bl); Wirestock (tr). **226 Getty Images:** Moment / K. D. Kirchmeier. **227 Getty Images:** Moment / Beata Whitehead (tr). **227-231 Dreamstime.com:** Bazuzzza (background). **228-229 Getty Images / iStock:** DrPAS (grey crows x5, crowsx3). **228 Alamy Stock Photo:** The Picture Art Collection (bl). **229 Dorling Kindersley:** Gary Ombler / University of Aberdeen (bl). **Dreamstime.com:** Steve Byland (cra); Kingmaphotos (cr). **230 Alamy Stock Photo:** Mc Photo (tl); ZSSD / Minden Pictures (bc). **231 Alamy Stock Photo:** LMA / AW (tr). **Bridgeman Images:** © Russell-Cotes Art Gallery (b). **232-235 Getty Images / iStock:** DigitalVision Vectors / GeorgePeters (background). **232 naturepl.com:** Marie Read (t). **233 naturepl.com:** Andres M. Dominguez (br). **234 Alamy Stock Photo:** Agami / Saverio Gatto (bl); The Natural History Museum (tl). **Dorling Kindersley:** E. J. Peiker (br). **Dreamstime.com:** Agami Photo Agency (bc). **235 naturepl.com:** Guy Edwardes (t); Markus Varesvuo (b). **236-237 Alamy Stock Photo:** Iconographic Archive (background). **236 123RF.com:** Michael Lane (tr). **Bridgeman Images:** Look and Learn. **Shutterstock.com:** EcoPrint (cr). **237 Alamy Stock Photo:** blickwinkel / AGAMI / D. Occhiato (br). **238-239 Dreamstime.com:** Markusmayer1. **238 Dreamstime.com:** Ndp (cla). **SuperStock:** © Michael S. Nolan / age fotostock (bl). **239 Bridgeman Images:** Photo © Brooklyn Museum / Gift of Mr. and Mrs. Peter P. Pessutti (r). **Shutterstock.com:** Manu M Nair (tl). **240 Getty Images / iStock:** DigitalVision Vectors / clu. **240-242 Getty Images / iStock:** DigitalVision Vectors / GeorgePeters (background). **241 The Metropolitan Museum of Art:** Gift of Estate of Samuel Isham, 1914 (br). **Shutterstock.com:** gergosz (tr). **242 Dorling Kindersley:** Mike Lane (br). **Dreamstime.com:** Martin Pelanek (bc). **naturepl.com:** Jussi Murtosaari (tl). **Shutterstock.com:** Agami Photo Agency (bl). **243 Alamy Stock Photo:** Jess Findlay / BIA / Minden Pictures (tr). **Getty Images / iStock:** DigitalVision Vectors / ilbusca. **naturepl.com:** Markus Varesvuo (crb). **244 Dreamstime.com:** Jozef Sedmak (tl). **244-245 Getty Images / iStock:** Milan Krasula (tr). **Shutterstock.com:** AlinArt (Background). **245 Dorling Kindersley:** Chris Gomersall (br). **Dreamstime.com:** Agami Photo Agency (bc); Caglar Gungor (bl). **246 Alamy Stock Photo:** FLPA (bl). **Dreamstime.com:** Mircea Bezergheanu (clb); Feathercollector (cl). **Shutterstock.com:** AlinArt; SarahLou Photography (tc). **247 Alamy Stock Photo:** ARTGEN. **248-249 Dreamstime.com:** Artinblackink. **248 Alamy Stock Photo:** Natalia Kuzmina. **249 Alamy Stock Photo:** steeve-x-art (br). **Getty Images:** De Agostini / DEA / A. DAGLI ORTI (cra). **250 Alamy Stock Photo:** Reading Room 2020 (tl). **Dreamstime.com:** Artinblackink; Benjaminboeckle (bl); Volodymyr Kucherenko (br). **Getty Images / iStock:** aaprophoto (bc). **naturepl.com:** Alan Murphy (crb). **251 Alamy Stock Photo:** All Canada Photos / Tim Zurowski (tl). **Getty Images:** Hulton Archive / Heritage Art / Heritage Images (tr). **252-253 Keith Williams. 254 Alamy Stock Photo:** Anup Shah / Minden Pictures (bl). **Dreamstime.com:** Mikelane45 (tl); Amit Rane (cl). **Getty Images:** Costfoto / Future Publishing (r). **254-255 Shutterstock.com:** Morphart Creation. **255 Alamy Stock Photo:** DBI Studio (br); J. Forbes (tc). **256 Alamy Stock Photo:** CPA Media Pte Ltd / Pictures From History (tr). **256-257 Alamy Stock Photo:** Danita Delimont (b). **Getty Images / iStock:** serkanmutan. **257 Dreamstime.com:** Glenn Nagel (cr); Thawats (crb). **258-259 James Crombie. 260 The Cleveland Museum Of Art:** Gift of Mrs. Henry Chisholm 1937.696. **260-265 Getty Images / iStock:** DigitalVision Vectors / duncan1890 (background). **261 naturepl.com:** Donald M. Jones (tc). **262 Alamy Stock Photo:** AGAMI Photo Agency / Brian E. Small (bl); David Tipling Photo Library (tc). **263 Alamy Stock Photo:** Wim Weenink / NiS / Minden Pictures. **264 Alamy Stock Photo:** Hira Punjabi (tr); Dave Watts (br). **naturepl.com:** Rob Drummond / BIA (tl); Ashish & Shanthi Chandola (tc). **Shutterstock.com:** Sergey Kohl (cra). **265 Alamy Stock Photo:** The Natural History Museum (tl). **266-267 The Metropolitan Museum of Art:** Purchase, Friends of Islamic Art Gifts, 2012. **268 Alamy Stock Photo:** phototrip (ca). **268-269 Getty Images / iStock:** benoitb. **269 Alamy Stock Photo:** Natural History Museum, London (ca). **270-271 Getty Images / iStock:** benoitb. **270 Photo © The Maas Gallery, London / Bridgeman Images:** Fitzgerald, John Anster (ca). **271 Dorling Kindersley:** Chris Gomersall (cr). **Dreamstime.com:** Xuying1975 (tr). **Shutterstock.com:** Independent birds (cb). **272 Dreamstime.com:** Stef De Rijk (bl); Jinfeng Zhang (cl). **Getty Images:** Sepia Times / Universal Images Group. **272-273 Alamy Stock Photo:** Quagga Media (Background). **273 Getty Images / iStock:** AlbyDeTweede (br). **Nicholas Kanakis:** (tr). **274-275 Terje Kolaas. 276-277 Getty Images / iStock:** Alisa Pravotorova. **276 Getty Images:** De Agostini / DEA / G. DAGLI ORTI. **277 Alamy Stock Photo:** AGAMI Photo Agency / Alex Vargas (cra). **Dreamstime.com:** Imogen Warren (cr). **naturepl.com:** Ch'ien Lee (tr). **278-279 Getty Images / iStock:** Hein Nouwens. **278 naturepl.com:** Richard Du Toit (b). **279 Dreamstime.com:** Assoonas (bl); Forest71 (tr); Bhalchandra Pujari (bc); Supaluk Payungwong (br). **280-281 Shutterstock.com:** Dn Br. **280 Getty Images:** Photodisc / Manoj Shah. **281 Alamy Stock Photo:** Francois Loubser (br). **Bridgeman Images:** Photo © NPL – DeA Picture Library (tc). **Dreamstime.com:** Artushfoto (tr); Mikelane45 (cra). **282-283 Getty Images / iStock:** Hein Nouwens (background). **Shutterstock.com:** COULANGES (ca). **283 Alamy Stock Photo:** Florilegius (bc). **Dreamstime.com:** Bespaliy (br); Kevin Gillot (tr); Gene Zhang (cr). **284-285 Dreamstime.com:** Ecophoto (b). **284 The Metropolitan Museum of Art:** Rogers Fund, 1913 (tr). **285 Alamy Stock Photo:** AGAMI Photo Agency / Ralph Martin (br); World History Archive (tr). **Dreamstime.com:** Pascal Halder (cra); Wirestock (cr). **286-287 Alamy Stock Photo:** Authentic-Originals. **286 Alamy Stock Photo:** Scott Leslie / Minden Pictures (cr). **Yale University Art Gallery:** Gift of Harrison F. Bassett in memory of his wife Elizabeth Ives Bassett and her brother Arthur Noble Brown (bc). **287 Alamy Stock Photo:** MET / BOT (ca). **288-289 Alamy Stock Photo:** Authentic-Originals. **288 Alamy Stock Photo:** The Natural History Museum. **289 Dreamstime.com:** Agami Photo Agency (cr); Mikelane45 (tr); Menno67 (br). **Science Museum Group:** Museum of Science & Industry (tc). **Shutterstock.com:** Martin Pelanek (cb). **290 Getty Images:** Smith Collection / Gado. **290-291 Shutterstock.com:** AlinArt (background). **291 Alamy Stock Photo:** Ralph Martin / AGAMI Photo Agency (t). **292 Alamy Stock Photo:** Ralph Martin / AGAMI Photo Agency (tc, bl). **naturepl.com:** Andy Sands (cra). **Shutterstock.com:** AlinArt (background); Dr Ajay Kumar Singh (crb). **293 Alamy Stock Photo:** imageBROKER / Gianpiero Ferrari (t). **© The Trustees of the British Museum. All rights reserved:** (crb). **294 Alamy Stock Photo:** Bill Gorum (r). **Dreamstime.com:** Kerry Hargrove (clb). **Shutterstock.com:** Geraldo Morais (cl). **295 Alamy Stock Photo:** The Natural History Museum (r). **Shutterstock.com:** AlinArt. **296 naturepl.com:** Alan Murphy / BIA (b). **296-297 Shutterstock.com:** AlinArt (background). **297 Alamy Stock Photo:** Florilegius (cb); Bill Gozansky (tc). **Dreamstime.com:** NatmacStock (cr). **298-299 Dreamstime.com:** Patrick Guenette (background). **298 Alamy Stock Photo:** Glenn Bartley / All Canada Photos (tr); Gabbro (br). **SuperStock:** Urbach, James (cra). **299 Getty Images:** Jeff R Clow. **300-301 Itamar Campos. 302-303 Shutterstock.com:** Olga Korneeva (background). **302 Alamy Stock Photo:** Ignacio Yufera / Biosphoto (cra). **303 Alamy Stock Photo:** Glenn Bartley / BIA / Minden Pictures (crb); Jacob S. Spendelow / BIA / Minden Pictures (br). **Getty Images:** Sepia Times / Universal Images Group Editorial (l). **304-305 Shutterstock.com:** Olga Korneeva (background). **304 naturepl.com:** Phil Savoie (l). **305 Alamy Stock Photo:** Rolf Nussbaumer (clb). **Dreamstime.com:** Agami Photo Agency (tr)

All other images © Dorling Kindersley Limited